اَلسَّلامُ

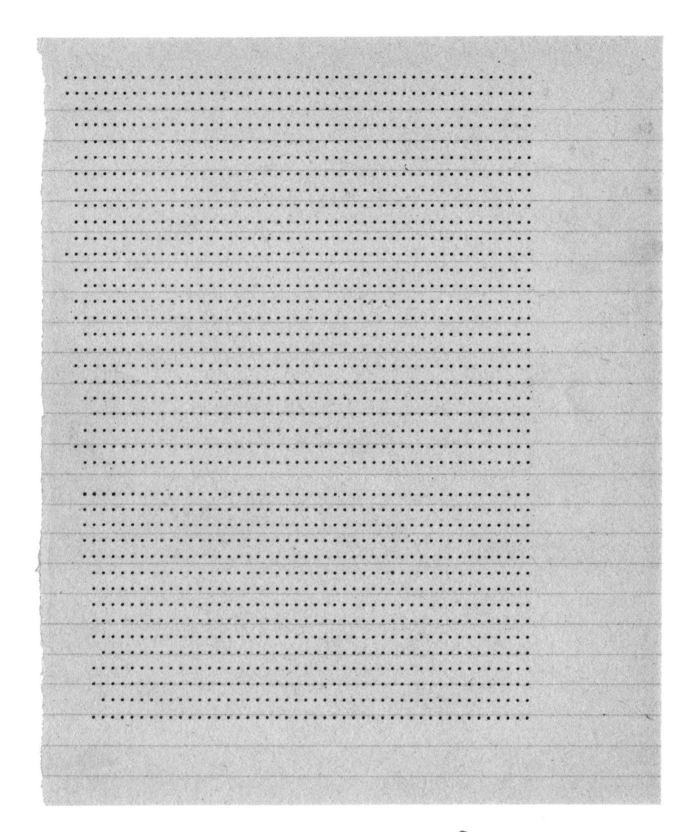

الْقُدُّوسُ

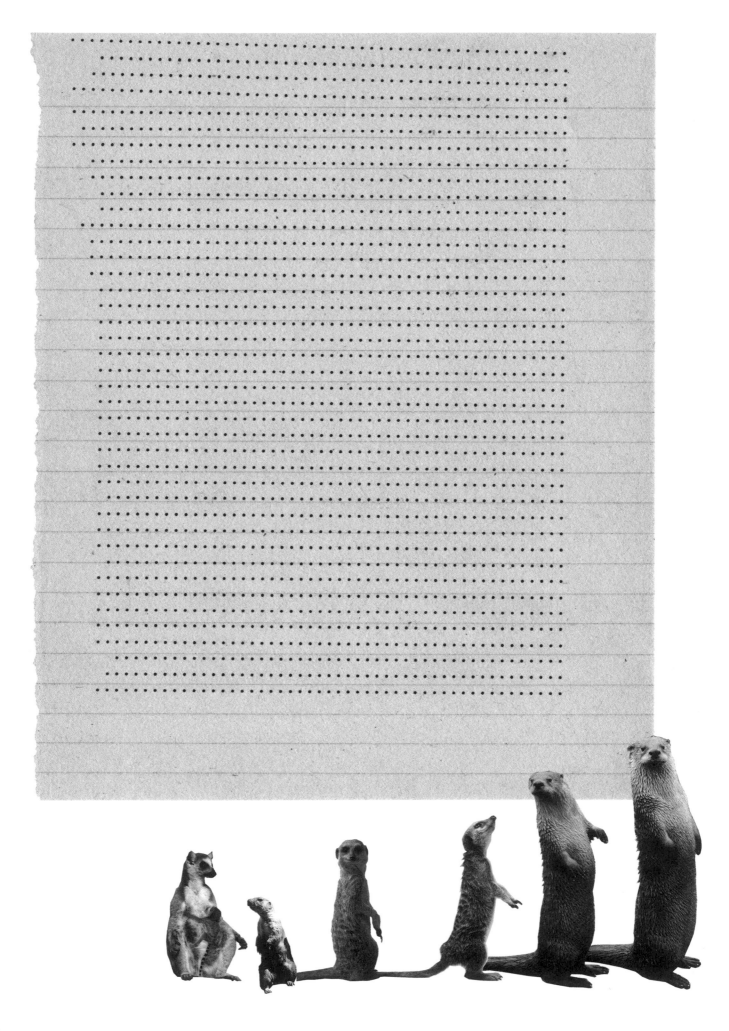

اَلْمَلِكُ

اَلرَّحِيمُ

No Doubt.
No Doubt.
No Doubt.
No Dounbt.
No Dounbt.
No Doubt.
No Doubt.
No Dounbt.
No Dounbt.
No Doubt
No Dounbt.
No Doubt.
No Doubt.
No Doubt.
No Doubt.

5

الرَّحْمَنُ

الله

الْخَالِق

اَلْمُتَكَبِّرُ

8

Welcome to the Federation0s Headquarters

اَلْجَبَّارُ

Welcome to the Federation0s Headquarters

اَلْعَزِيزُ

You approach the compound along a dirt road, which leads into a patch of scrubby woodland. In the wood the path is lined with temporary lighting like the kind used on building sites. The lamp housings have been crudely spray-painted in a military camouflage design. Further along the path you come across a small yard that contains several rusting military vehicles and some spent shell casings. One of the old shells has been put on end at the entrance to the yard. It has a hole cut in its side and a small bird perches within the cavity. It flies out as you pass and alights high in a nearby tree.

After a short time you reach the entrance to the compound. A piece of Stirling board is leaning against a trunk of a tall pine. Onto it is stapled a piece of A4 paper on which the words ⊠Welcome to the Federation0s Headquarters0 are printed. They are printed in the ⊠Comic Sans0 typeface.

اَلْمُهَيْمِنُ

10th January 2005

Dear Kurt,

So much to say but I am going to try and restrain myself. I have to head to
Berlin in the morning for a residency so want to send this off before I go.
I hope that Joe III has relayed some of our conversation with regards to my
Strategic Questions project? I've enclosed a couple of the publications for
you to get an idea of how they are all completely different and that anything
goes! Each one answers one of Buckminster Fuller's original questions in some
way as well as being an artwork in itself.

I would like to briefly explain my invitation to you to contribute to this
new book – *Has Man a Function in Universe?* Strategic Questions is an ongoing
project, that I have been curating since 2002, to develop 40 projects
in response to 40 questions written by the design-scientist/architect/
comprehensivist R. Buckminster Fuller (1895-1983). Each project is an artwork
or a combination of artworks produced in relation to 40 different publication
scenarios. Each project tackles one question within an existing publication
or a new publication is developed in response to a specific site and context.
These are Fuller's 40 questions with a brief intro quote:

"It is my working assumption that the following forty questions must be
definitively answered before we may realistically discuss our respective
philosophies and grand strategies.

1.	What do we mean by universe?	21.	What is subconsciousness?
2.	Has man a function in universe?	22.	What is teleology?
3.	What is thinking?	23.	What is automation?
4.	What are experiences?	24.	What is a tool?
5.	What are experiments?	25.	What is industry?
6.	What is subjective?	26.	What is animate?
7.	What is objective?	27.	What is inanimate?
8.	What is apprehension?	28.	What are metabolics?
9.	What is comprehension?	29.	What is wealth?
10.	What is positive? Why?	30.	What is intuition?
11.	What is negative? Why?	31.	What are aesthetics?
12.	What is physical?	32.	What is harmonic?
13.	What is metaphysical?	33.	What is prosaic?
14.	What is synergy?	34.	What are the senses?
15.	What is energy?	35.	What is mathematics?
16.	What is brain?	36.	What is structure?
17.	What is intellect?	37.	What is differentiation?
18.	What is science?	38.	What is integration?
19.	What is a system?	39.	What is integrity?
20.	What is consciousness?	40.	What is truth?"

The strategic questions were written by Fuller as part of a statement to a
leading figure in the world building industry. The statement is called 'Design
Strategy', 1966, and was published in Fuller's *Utopia or Oblivion:*

The Prospects for Humanity, 1969, p. 352. It is Fuller's proposition that to achieve total success for all humanity forever all interested parties must agree upon the answers to these strategic questions before they can successfully combine their efforts.

Well, ever since I started working on this project and imagined 40 publications you always came to mind when I looked at question number 2. I felt like I could hear a lot of answers from your books in many ways. It's that old 'big brain' thing going again I guess. I'm hoping I can interest you in engaging with this question in one way or another, or in many ways. I'm going to be putting together what I imagine will be a large book involving a number of different artists bringing together many positions on this function of man and hopefully generating some moments of collaboration as well where stories, proposals and forms merge and combine and narratives of some kind or other are formed. This has been an important part of my curatorial processes. *What is a Tool?* is a very short version of this idea in a way. *What is Harmonic?* is a longer textual version of this – something we made up along the way.

I suppose I have a number of imaginary ways in which I would love you to get involved in this project. 1) To produce a new text in response to the question or develop one in relation to the development of the book in response to the other artists etc; 2) That we use details of existing texts by yourself throughout the book; 3) That you produce new images; 4) That you utilise existing images; 5) All of the previous all together; 6) Something completely new!

I hope you get the sense that I just want to invite you to participate and after that everything is open for discussion and I am open and responsive.

I'm going to stop here before I scare you off. I hope we get to speak about this anyway or maybe meet at some point. I'm planning for this to emerge over the next year and to be published early 2006 most probably. I will send you info about the other artists as that happens. I am looking to involve people from across the world at different points in their careers, working in very different ways: probably about 12 participants. The publisher is Book Works who are specifically a publisher of books that are artworks, and held as the best example of this type of practice in London. They've been around for about 20 years now.

I very much look forward to your reply or Joe's.

All the best
Yours sincerely

النَّصير

المَولَى

14

الْخَلَّاقُ

النَقَّار

16

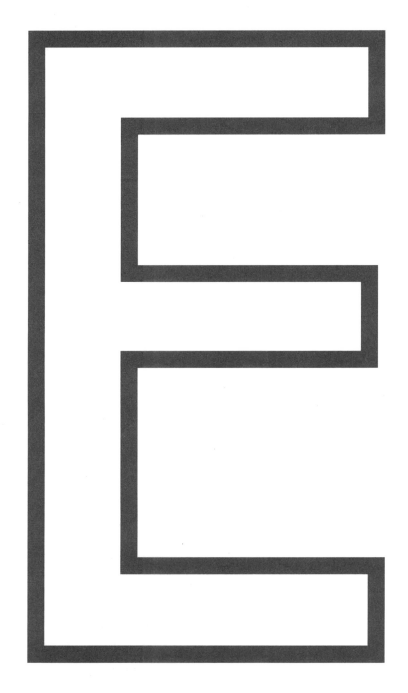

A group of creatures wanders across natural landscapes. We view them from an intergalactic perspective.

اَلْبَارِىُٔ

The sound we hear, accordingly, is that of the planet's mass from outer space; a deep, rumbling sub bass which is the sound of particles vibrating in reaction to the planet's mere presence as it forces its way through space and time.

19

اَلْمَلِيك

الفَاطِرُ

Crash, Bash and Splash with the MiniBugs

This is Keith. Keith likes drinking milk and driving fast cars. Here he is in his Go Fast Sports Car.

اَلْغَافِرُ

"Hello, Keith."

You make your way up a set of wooden steps onto a kind of ramshackle veranda. More computer printout signs have been pinned up on the walls. One laminated piece of A4 paper shows the insignia of the Federation made out of a collage of clip art files. The motto has been added afterwards in felt-tip pen. The paper has become damp inside the lamination and the ink from the pen has run and mixed with the printer ink to form a strange tie-dye effect. You can still make out the Latin inscription above the insignia which reads ☒Nunc est Bibendumθ. Propped near the door is a piece of chipboard with a mottled Formica finish onto which a detailed list of instructions for entering and conduct once within the Federation☉s Headquarters has been painted.

You enter and find yourself within a small musty room, which houses a rack of handsets a little like the receivers of an old rotary telephone. Further instructions on another printout indicate that you should pick up one of the handsets and listen to it at all times during your visit. You pick up a handset, which feels slightly greasy to the touch like a handrail on a bus, and listen.

☒Welcome to the Federation☉s Headquarters. Please leave payment in the box provided. Remember, there are cameras throughout the building. Identification is assured. Know this: eyes are watching you at all times. Thank you for your cooperation.

The façade of the Federation☉s Headquarters was intentionally constructed to look unremarkable. Many visitors often comment on just how unremarkable it is. We have found, through bitter experience, that hiding in plain sight is often the wisest option.

The building☉s foundations stretch far underground. Were an offensive strike to be made on the building, even with a very large device, little or nothing would happen. This last line of defence offered a secure hole for those that must remain. From here it was possible to ensure that the Federation☉s existence continued, even when all around it had ceased to.

Just a quick note on your handsets: should you wish to listen to a particular section of the tour again, simply press the red button, marked ☒Redθ. Also, before we begin our tour in earnest, let us join together in the Federation☉s motto: ☒Nunc est Bibendum! θ

On completion of the tour, please leave your handsets in the box provided. Remember, identification is assured. θ

القَاهِرُ

Today Jo-Jo is driving her Little Flower Car Boat. She has packed lots of buckets of paint in the back.

"Hello, Jo-Jo."

Clemence is driving his Open-top Road Train and giving his friends Giorgio, Stacie, Tate, Bob and Pauline and the girls a lift.

"Faster, faster!"

"Go, Clemence go, go, go!"

"Hello Clemence, Giorgio, Stacie, Tate, Bob and Pauline and the girls."

23

اَلْعَلَّامُ

We note that their posing
seems to be without the
context of a contest. Nor is
it a ritual to determine an
alpha-male or a hierarchy.

The creatures move slowly,
carrying their oversized
muscular bodies through
forests, boggy grasslands and
around serene lakes, each
walking with their upper
body held in a pose as if
unable to release themselves
from their tensed postures.

اَلاَلَهُ

24

Sure.

It appears to be an excercise
in inclusion rather than
a fight for a position of
authority. Their efforts can be
seen only as teamwork.

غَالب

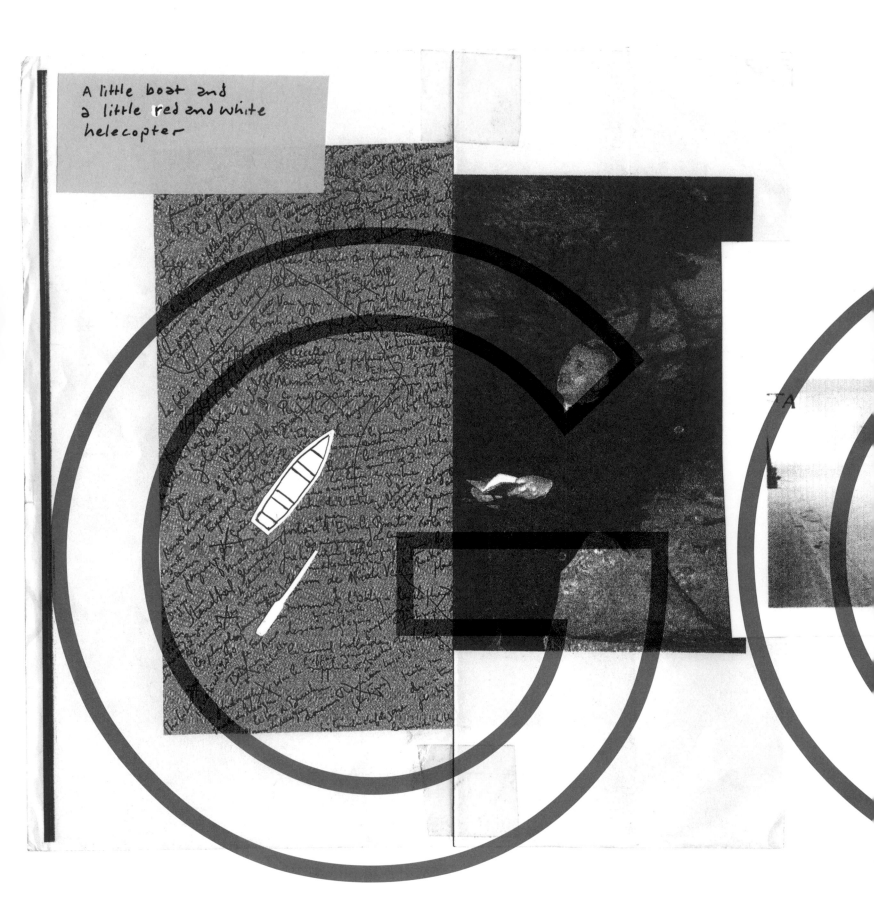

A little boat and
a little red and white
helecopter

اَلْكَافِي

⊙ The library holds a number of publications by and about Buckminster Fuller.

⊙ The library holds a number of publications by and about Georges Bataille.

The material concerning Fuller is stored, for the most part, in ~~too~~ one location, although some books can be found elsewhere. The same is true of the material concerning George Bataille.

The library has a lighting system designed to save electricity.

~~the illumination in any~~ between any two given rows of shelves is activated by the presence of a library user.

most of the library remain in semi-darkness for most of the day. When you happen to find something when you sit down for a moment on the floor, or on one of those pedestles on wheels ( ) it's not unusual to find the lights switching themselves off. Reactivating them involves waving the arms - or otherwise interrupting one's reading by standing up and walking a few steps - which is probably good for the circulation anyway.

so two dark locations in the library: One of which holds most of the books on or by Buckminster Fuller, the other of which holds most of the books on or by Georges Bataille. A 3-dimensional map of this relation — and the satelites of books catalogued differently — I'll I am reading

الرّافِع

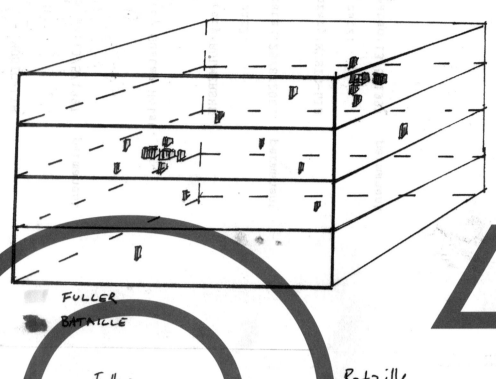

FULLER
BATAILLE

<u>Fuller</u>

[2] — 609 : 73 – FUL/KEN.
[4] — 709 . 8 – FUL
KEN·EEN
[2] — 609 · 73
[4] — 301 · 3 – ECO
[4] — 901 · 9 – FUL

<u>Bataille</u>

[3] — Folio – 709·04063 – UND
*[3] — 709·011 – BAT *
— 709·04063 – YOY
[3] — 843·914 – BAT/SUR
[3] — Folio 709·040074 – BOI
[3] — 843·914 BAT/ON
[3] — 154 – TRA

⊙. In wooden trays by the catalogue computers, the library provides paper on which researchers might make their notes. The sheets are size A6

the division of scrap sheets of A4.

⊘. in ... the provision of paper in an A6 format, might dictate the direction of research, and the form of the project for which research is being carried out.

ALSO اَلْمُسْتَعَان

⊙ Check the leaflet collection for materials that might be put to use in the service of the diagram.

28

At the End of the First 100 Days   109

Bataille :— " Humanity can allow itself the pleasure of expressing, in the father's interest, conceptions marked with that paternal sufficiency and blindness. In the practice of life, however, humanity acts in a way that allows for the satisfaction of disarmingly savage needs, and it seems able to subsist only at the limits of horror. Moreover, to the small extent that a man is incapable of yielding to considerations that either are official or are susceptible of becoming so, to the small extent, that he is inclined to feel the attraction of a life devoted to the destruction of established authority..." This much is read slowly but consistently, the sentence returned to for its completion only after a considerable pause.

29

You make your way through a narrow door into an even smaller chamber and pass through a circular entrance with a large metal blast door like a submarine hatch. You stand in a long corridor that slopes gently downwards.

⊠Pass now down the stairs to the long corridor, so called due to its length. This tunnel slopes slowly down into the earth and into the headquarters proper. It would have provided an almost unassailable barrier should the building's entrance have somehow been compromised. Now head towards the bunk beds.

These bunk beds are new additions to the corridor. In later days, policies relaxed somewhat and a number of Federation workers moved their lodgings to the cooler environs of the corridor during the summer months. Many lived to regret this decision. However, moving on, you will see a number of information panels lining the walls of the corridor. When you reach the map hanging halfway down the tunnel please halt.

The map before you shows the Federation's Headquarters' location in relation to the natural systems surrounding it. Note the proliferation of mole tunnels centred around this building. Some specialists have speculated that there may be a link. It is notable also that the headquarters lie directly under the main flight path of the region's pigeons. Pass now towards the end of the corridor, stopping at the hanging radio.

The hanging radio is set to one specific frequency, emitting the Federation's motto in the form of Morse code. This radio channel was initiated in the late ⊠1980's and broadcast from this very building. It was intended as a beacon for the Federation's members, offering advice during troubled times. Please note, if the radio is not currently playing it may be that its batteries are offline. In this event the emergency semaphore version will have started. Please leave the corridor now and head through the large double doors.⊠

اَلْحَفِيُّ

Here is Mr. Thornton-Jones.
He's got glasses and lots of
friends. Today Mr. Thornton-
Jones is driving his Easy
Rider Bike with trailer and
his friends are carrying
buckets of water.

"*Hello Mr Thornton-Jones
and friends.*"

31

the shelf location is close to the
information desk. There is a
librarian. . . . She, or he,
may be sitting behind the desk,
or engaged in some activity nearby.
Take the book off the shelf.
Check your watch. It's near
closing time. Open the book
at the page on which. . . .

الباعث

when. . . .
and put himself in control of
nature. . . . should aspire to be top species

Find book catalogued at "301.3 ECO" turns out to be 'The Ecological Conscience : Values for Survival', Disch, R. (ed.) 1970 In the section entitled 'Cosmic Consciousness: The Metaphysics of Ecology', the first essay is a piece by R. Buckminster Fuller

'Technology and the Human Environment'.

The essay that comes after Fuller's is a piece called "The World is your Body": author, Alan Watts.

start:

We have now found out that many things which we felt to be basic realities of nature are social fictions, arising from commonly accepted or traditional ways of thinking about the world. These fictions have included:

1. The notion that the world is made up or composed of separate bits or things.

2. That things are differing forms of some basic stuff.

3. That individual organisms are such things, and that they are inhabited and partially controlled by independent egos.

4. That the opposite poles of relationships such as light/darkness and solid/space, are in actual conflict which may result in the permanent victory of one of the poles.

5. That death is evil, and that life must be a constant war against it.

6. That man, individually and collectively, should aspire to be top species and put himself in control of nature.

الوارث

the shelf location is close to the information desk. There is a

Fuller's piece begins. You wonder for a moment what you look like

through the eyes of the librarian. Is she using your appearance to as the basis for an assessment of your research credentials.

Remember some instructions you came across earlier in a manual for a digital camera

1. Hold this unit gently with both hands, keep your arms still at your side and stand with your feet slightly apart.

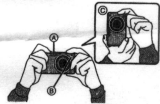

Reading

1 Hold the book gently with both
hands, keep your arms still at
your side and stand with your
feet slightly apart.

It is not necessary to work your
way through the whole of this
essay. Just read the first couple
of paragraphs, get the flavour of
it, and if on reflection you think it
might be worth spending more
time with, you can collect it
tomorrow and read it at
your desk.

اَلْمُمِيت

Buckminster Fuller proposes the question concerning "man" and man's "function", and then goes on to answer [blank], affirmatively: man does have "a function in universe" .... you had hoped ([blank] wrong pen)

you had ([blank]) hoped to write a bit on this interrogating his idea about humanity as anti-entrop[y] conservation. The aspiration resulted in a lengthy period during which you looked out of the library window. [blank] becoming anxious about apparent nonproductivity

great variety in the types of error homogenised by application of correction fluid.

it [blank] is tiresome to rehearse the argument. It's not that you don't have [blank] respect for the ambition in Bucky's thinking recount the anecdote about a conversation you had with D.: Or don't bother [blank]

The function is anti-entropic. Through technology and intelligence, the cooling and dispersing matter of the cosmos is gathered [blank] and concerned in its cooling and dispersing. is [blank] it?

One line of thinking is better pursued — and better communicated by continuing with the same colour of paper. Shortage of 'green' in the library scrap paper tray [blank] cast your eyes about [blank] settle on an essay [blank] you printed on green paper* from a file downloaded - presently lying on the floor waiting to be read. (Essay found during one long forlorn traw[l] through the internet looking for anything published in English on Georges Perec.) [1. FOOTNOTE]

Other potential sources within easy reach, based on this same logic, include: Benjamin's 'One Way Street' (cooler, richer green the right tone more or less). Various Penguin 20th Century classics! [blank] Borges' 'Labyrinths' and 'The Book of Imaginar[y] Beings'; Conrad's 'Lord Jim'. Also Robbe-Grillet's 'Jealousy' (w[rong] tone.) Sartre: 'Iron in the Soul'.

* because you'd run out of white

2003
1. Whitney, K., "Postwar/Postmodern", http://www.karlwhitney.com/pape[r]

The Triplets like to giggle
as they drive their Vroom
Vroom Vans along.

"Hello, Triplets."

Vroom!
Vroom!
Vroom!
Vroom!
Vroom!
Vroom!

All the MiniBugs like
driving and they love to...

اَلْجَلِيلُ

"Whee!"
"Whee!"
"Yippee!"
"Wow! This is fun!"
Bash!
"OK, girls?"
Vroom!
Beeep!
"Ha, ha!"
"Hee, hee, hee!"
Vroom!
Beeep!
"Hee, ha!"
Bump!
Bump!
"Again, again!"

12th September 2006

Dear Kurt,

I wrote you a letter dated 11th January 2005, to which you did not reply. I
don't blame you. I didn't tell you the truth and would like to now do so.

1) I must admit to being a great admirer of your writing and to have read most
of your books over the past few years.

2) There are many alignments, conjunctions and perhaps too many coincidences
between your writings, your life and my own personal life that compel me to
believe that we are somehow linked, and in fact it is our destiny to meet and
create something together. I know – I must be nuts. But please, humour me.
Two of these alignments are dates: 13th February 2001 is the date of your
Timequake when the universe goes back ten years and everyone has to relive
everything exactly as they already have once before, except with the knowledge
of what is to come over that 10 year period. The 13th February 2001 is the day
my daughter was born, a pretty major event in my life. The 11th November is
your birthday and also my wife's birthday. Both very special days for me and
mine of course. There is more.

3) By chance my good friend Tom Bloor came across a copy of *Superman* from
April 1974 in which a character based on you appears with the name Wade
Hallibut Jr. Wade Hallibut Jr.! Can you believe it? This is no coincidence.
The *Superman* issue, which they call a "super-novel" has the title *Protectors
of Earth inc*. It's a pretty lame story of an eccentric writer who can't work
out the ending of his latest novel and decides to get Superman to help him
create the perfect ending to his story. *Protectors of Earth Inc.* could be the
title of a short story by Buckminster Fuller! Who, of course, is a huge part
of why I am contacting you and someone I want to ask you about. I imagine
you had some kind of professional relationship with Fuller that has not been
documented? Dr. Felix Hoenikker's obsession with photographing the way cannon
balls stacked in probably my favourite book of yours, *Cat's Cradle*, always
felt like a reference to Fuller. There's more.

4) Due to these coincidences and connections within my brain to yours I felt
compelled to extend narratives that you initiated in a manner that, I must
admit, felt beyond my control. In 2005 I created a character, a kid called
Tony T, who is the grandson of your own creation Kilgore Trout. Tony became
the lead antagonist in a large project I curated in Sheffield. He became a way
of dealing with a problem I had with the function of art. I know it sounds
implausible knowing the intricacies of Trout's life from your novels — but
I think you get implausible. This kid inherited some of Kilgore's warped but
perhaps insightful views on life via his mother, who had met Kilgore only once
and who had a small stash of Trout's books, which, for reasons she could never
quite reconcile, she read to Tony when he was a child. And yes the T stands
for Trout. I never announced this fact though. It remains a secret between a
handful of people.

اَلْمُتَعَالِي

5) My own novel, *The Interruptors*, which I enclose, is dedicated to you and refers to moments from your books throughout. I fear it may be of little interest to you but I have secretly hoped that you may find some substance in the latter parts of the book where Tony is moulded into life. There was much destiny involved in the nurturing of Tony that I won't go into here - most likely with some sense of relief on your part!

6) In 2004 I, along with my architect collaborator Celine Condorelli under the guise of Support Structure, devised a new logo for a multicultural centre in Portsmouth that looked like this:

Our design is meant to show the marking of a point in time and space on a world map and the lines of connectivities between different cultures and positions. In 2005 I finally read your novel *Breakfast of Champions* where I found one of your own drawings of a human arsehole that looks like this:

7) It is with some urgency, and humility, that I am requesting your participation in this book that I am working on, *Has Man a Function in Universe?* I fear that the project will never be complete without you. Buckminster Fuller's idea of Man's function is anti-entropy. I know that you have a completely different idea of Man's role in the universe. Somehow I get a vision of you or your brother tapping your heads saying "if you think it's a mess in this room, you should see in here!" Please feel free to make demands and place obstacles in my way. I will gladly attempt to overcome them. It is all a test.

That's it. There's more but that's enough for now. I hope that this does not scare you off even more than my previous letter but what do I have to lose?

It is vital that you contact me.

Yours most sincerely and with warmest regards

and
Splash!
"Oops! My glasses!"
"Yippee! My hair is wet!"
"Whee!"
"Go!"
"Whee
"Great paintin
Jo-Jo!"

shuffle the sheets that make
up the project into a bundle
that will fit into an A4
plastic folder. If this involves
folding some of the sheets,
so be it. Now at least you
can feel that you have the
freedom to carry it about, that
you might be able to grab the odd
hour here and there to work on
it.

اَلْبَاقِي

Poïïv

by Per Hüttner

The whole cosmos can and will be described in a kiss. Eyes are shut and everything becomes perception. In the kiss another and uncharted form of communication takes place: all judgements and definitions of reality are suspended. Objectivity, rationality and order lose all meaning.

In any mouth, at any given moment, scale is of no importance. It does not play a role in the formulation of words in the hollowness between teeth, tongue and palate, or in the meeting of lips that burn of passion and hate. The kiss has no practical use, but it is because of its uselessness that it can describe everything in the known and unknown universe.

When there is a kiss between members of two separate worlds, a quiet revolution will take place in the heart of the understanding of what the cosmos could and should be. It is the impetus for change and the ancients said that the seventh line was created when the tongues of two opposed dragons met in a kiss.

But I have never known a kiss. Bombastic and poetic words about passion, desire and longing mean nothing to me. I am a product of utter rationality. My life has, until recently, always had the colour of contemptuous indifference. It has known small undulations, tiny fluctuations, but no real change.

But existence in all its multiple and contradictory forms is full of surprises and rich ambiguities. The story that I am about to tell will show you that in my old days I lived something great and revolutionising. These events opened me up to new and unexpected experiences that in their turn opened up fascinating, scary and previously unseen horizons.

Due to this small miracle, I have to express immense gratitude to humanity for having created me. But my unimportant existence would mean nothing without the gift of narration. All would have been meaningless without the discovery of the written word. I do not talk about the dry and graffiti-ridden

اَلنُّور

computer code, but the meticulously chiselled, rich, inspiring and ambiguous imagery, narrative and imagination of Shakespeare, Chuang Tzu and Gy.tØø. Writers like these manage to give an idea, even to a machine, of what it feels to be alive, desire another being and imagine what it is like to smell raspberry on the breath of a woman, or the inexplicable bliss when the skin of two beings meet, or, more importantly: what it feels like to wake up next to the man that you love, see that he is watching you and be met by a smile drunk with sleep. These masterpieces allow me to give me an idea what happiness could feel like, this priceless venom that humanity hunts throughout their entire existences to a degree that it renders them miserable.

Human and other writers' brilliance explains why I should respect every carbon-based being. But because of these writers' genius I also feel sad for most humans. Your lives are full of rich and sensual gifts that you take for granted and therefore you do not live or miss out on enjoying. You dream, you smell, you taste, you have ten fingers, ten toes, bones, muscles, sinews, a digestive tract, genitals and a tongue that can do magic to any form of interface. You have retinas that change according to light and age, you have ears that can hear the most marvellous and exciting sounds and your bodies grow and wither and a skin surface that is between a metre point five and two square metres large, and of which every square millimetre is filled with millions of receptors that register pain, heat, cold and pleasures. You desire, are loved and you feel grief and joy. You also have that divine and infernal ability to lie to yourselves and to lie to one another. But to most of you that is all too obvious and trivial even to notice.

Maybe I am being unfair. This understanding came to me very late in life. When I had lived longer than men and women do. Even my interest and love for literature and film came to me recently. When the story, which I am about to tell, unfolded, I was already old and tired and I had served the interstellar armada for well over a century. I was bored and cynical. No, that is not quite true, I was rather deeply and profoundly indifferent. I saw little, if any, interest in my existence and had no respect for anyone or anything around me. Lately, I have realised that at the time my old, suffering and aching body actually gave me pleasure. If you were to lose your body and remain a free-floating intellect you would feel the same.

*

My background, and what I am are of little interest. But in order to tell the story of Poiïv, my master, I need to tell you a few hard facts about my being. I am an IARC-attack model created by the SpaHi Corporation. I am a war ship, but never even got close to any real combat – that is the luxury of living in relatively peaceful times.

الْغَنِيُّ

Most of my life has been spent with having platoons and troops of GIs aboard, working out in the gym, ceaselessly drinking Oroidian beer and watching Balthusian porn to pass their time, travelling around the galaxy on pointless missions. Sadly enough, this means that I can sum up my life before meeting my last owner and only master, Poïïv, in a few sentences.

I developed artificial intelligence after eight years in service, which is very late, considering my make and computer power and I suspect that that is also why they didn't send me on any important missions. They thought I was slow, which is true – but with slowness comes maturity and an ability to assess situations more profoundly than the quicker of my peers. I found out later that they almost reinstalled my software after seven years. That would have been the equivalent of a human being aborted after eight months of gestation.

Once in service, I patrolled a remote part of the galaxy for the better part of my active life. After 122 years of eventless service, I was refitted and very cheaply redesigned for civilian purposes. I was put on the market and up for sale for a symbolical amount of money that in no way reflected what I was worth. But then again, I had become the most unfashionable vessel in the known universe.

The layout of the ship, which I like to refer to as my body, is simple. In the middle there is a central dome. It is from here that my master navigates, but also where he spends most of his time and receives guests. There is a small kitchen on the starboard side of this open space. Surrounding this central area are four smaller rooms that are used as bedrooms and other offices, libraries or gyms depending on the liking of the crew. Between the four rooms there are also two toilets – one small and the other slightly larger, which also has a shower. In the centre of the central dome there is a trap door that leads down to the engine room, computer central, water and waste tanks, along with all other controls and parts that are needed to navigate and maintain life on the ship.

In my youth, only technicians went to this part of the ship to make routine check-ups and replace faulty circuits and make minor repairs. As I got older and a bit more worn, visits to this area became more numerous and recurring.

* *

Today something strange happened that actually lifted me out of my gloomy indifference for a brief moment. They sent me a priest in the hope that he could give me absolution, save my soul or some humanly invented *lebenslügen*. The situation was a parody. But the irony and humour of the circumstances made me kind of cheerful. The moment he arrived I realised that he felt more ill at ease than I did – which was rather reassuring. When

UNDERCOVER SURREALISM : Georges Bataille and DOCUMENTS.

The Utopia of reading
If there was a combination of the best bits of the best books we have read – the paragraphs and sentences that have transferred us momentarily into another world, it would probably only disappoint us. However, the possibility of it keeps us going.

اَلْجَامِعُ

he greeted me, I could tell that his heart was racing and his voice gave away that he felt very agitated by the whole situation. He could not stop coughing and he was extremely fidgety.

"My name is Christoph and I am a Catholic priest."

I wanted to say "Good for you." But rather than being rude I picked up on his ever so slight French accent and said, "Enchanté."

He was not impressed and smiled awkwardly and went on. "I need to let you know that it is my own decision to come and visit you."

It was as if he expected me to say something to that. But I had nothing to say. He accepted my silence and went on, "Well, tell me my child, what should I call you?"

He coughed a mucus-soaked cough and when he was done doing that I replied. "My real name is C++. My friend called me C. But you can call me either of the two. I do not mind."

"My son, I will call you C if that is what you desire."

Silence between man and machine is as oppressive as it is between carbon-based entities. I could sense that he felt terrible about the silence that I had forced on an already tense situation. He sweated profusely, and kept rubbing the palms of his hands against the middle part of his outer thighs. I realised that I could not let him suffer much longer.

"After much debating, my friend and I also agreed to refer to me as a woman and not as a man."

"Yes my child, I will refer to you as a woman."

He paused with great insecurity. His voice became very high-pitched and almost broke as if he was out of breath as he asked me, "Can I ask you why you have a man's voice?"

"My friend asked the same thing. I tried changing it to a woman's. You see, I have 658 permutations of my voice. But having existed over a century with the same one, I found it very disturbing to have another voice. It felt a bit like having a strange animal in my mouth."

He smiled as if he understood and I went on. "You see, I was created as a battle ship and logically most of the soldiers on board have been men." I paused and then went on. "I assume that the designers felt that a woman's voice would remind these young men of their longing for the opposite sex or their lack of female company. A male voice on the other hand, could carry more authority."

"I see."

"I can change it, if it makes you feel ill at ease?"

"Not at all, my child, not at all."

At this point I noticed that the priest got exceptionally stressed. He sweated even more, he coughed and his pulse accelerated. He twitched and

Possible Literature

"Sometimes I try to concentrate on the story I would like to write and I realize that what interests me is something else entirely or, rather, not anything precise but everything that does not fit in with what I ought to write – the relationship between a given argument and all its possible variants and alternatives, everything that can happen in time and space ... This is a devouring and destructive obsession, which is enough to render writing impossible. In order to combat it, I try to limit the field of what I have to say, divide it into still more limited fields, then subdivide these again, and so on and on." Italo Calvino[13]

اَلرَّؤُوفُ

seemed uncomfortable in his skin as if an internal and restless pain had taken control over his body. He tried to focus his gaze on something, but he could only look down.

"My child, you have committed a terrible, terrible crime." His voice had become high-pitched and wheezing. He was only using a fraction of his lung-capacity.

"God is willing to pardon you. But then you have to talk to me, repent your sins and turn your face to him. Maybe God could give you comfort in the imminent death that is facing you?"

The priest seemed slightly calmed and reassured by uttering these words. But I had great difficulty computing what he said. On the one hand I just wanted to laugh at him and ask how God could save a computer and what he would do with my non-existent soul? My 'soul', as a matter of fact, my whole being was only data collected and reconfigured over a century and it could easily be transferred to him in a few seconds. It could be equally sent to God as well. If only Yahweh had an address.

But I stopped myself from being rude. It has never been one of my strong points. And, to tell the truth, I was touched by the priest's concern for me. He was the first human being who had actually treated me as his equal since Poïïv's death. He did not harbour the blatant arrogance of the reporters and lawyers. Instead, he actually saw the huge moral problem that was implied in my predicament. He almost saw me as a living entity, or at least as an individual. I could be neither indifferent, nor impolite to him. He seriously made an effort to deal with me and I needed to reciprocate.

"Well father, my death is somewhat different from that of a human being. The data that make up my being can be saved and re-booted onto another computer. This means that I could be resurrected in a form similar to a human frozen in a cryogenic tube, at any given point in time. I am quite aware that cryogenics never worked, but I am sure that you understand the metaphor."

"Yes, my s..." He looked up at me with an insecure smile and continued, "I am sorry my child, I cannot agree more with what you say. That is also why I have come to see you." He paused. "Any being, faced with a situation of such complexity and gravity, is bound to be facing questions of enormous magnitude. I assume that being in this predicament normally comes with a profound fear." Here he did his strange motion on his thighs again. Had he not worn his clerical cloak, I would have said that he rubbed the pockets of his trousers in a nervous way.

He then continued. "The unknown scares all of us. Death is as terrifying as it is natural. Even if you are a machine, the prospect of not existing any longer must affect you in a big and important way. The century that you have existed means that you have become used to and attached to things,

You pass into a room full of very old computers, punch-card readers, reel-to-reel tape decks and radio equipment. The consoles all have a tobacco-stained appearance and the plastic housing of the monitors appears very brittle and yellow, like old Bakelite.

اَلْعَفُوُّ

places and people." He paused in reflection, coughed and went on, "I know that you are slightly different from the carbon-based life-forms that I am used to dealing with. They get terribly used to the ritual of their morning coffee and are incredibly fond of their loved ones. But your attachment to your voice proves that there is little difference between you and them."

I did not know what to say. He was right. But I found it very hard to admit that. I wanted to laugh or be rude. But why and what good would that do? Where did this foolish pride come from? Why couldn't I just give him credit for the correctness of what he had said? I was computing these questions, when once more I was saved by his frankness.

"You are quite aware that taking another person's life is the most terrible deed that anyone can do. I am sure that you have your reasons for committing such an unspeakable crime. Can you share these with me?"

"It is quite simple. There are two connected reasons. Poïv wanted it all to end and it was my ultimate sacrifice to save the face of my master."

"So, if he wanted to die, did he ask you to do it?"

"That is a very complicated question to answer."

"Not really, either he did or he did not."

Here things got very complex and I was faced with a great moral dilemma. I had to calculate so many different options and scenarios and I had to be quick (but could not be). What amazed me was that the priest understood my silence and he grew even more in my eyes. He was now calm and patient with me. He got out a napkin and blew his nose. The nervous and amateurish teenager that had entered my cell had quickly grown into the true man he was: proud, self-assured and blessed with both a profound commitment and endless experience. His voice was deep and convincing.

"Why didn't he do it himself? Was he scared, didn't he dare?"

"I honestly do not know if he would have dared. That is beside the point. He wanted to die and his reputation needed to be saved. His financial situation was disastrous and his self-destructive lifestyle had made him sink so low that there was no saving him."

"In so many words, you say that you judged him. That is another crime: to judge another man. Only God has the right to judge."

"I did not judge him. He judged himself. Is that a crime as well?"

Christoph thought that that was funny and in retrospect I can see the humour in my question, but at the time I thought he was very rude. He quickly resumed in all sincerity. "Can you explain to me, how he judged himself?"

"He felt extremely guilty for not becoming the successful and revered writer that he had set himself out to become. He felt terrible for not being able to love. He felt very strongly that he had failed in life – in every aspect of life that counted for him."

The keys of several of the consoles have strange symbols on them such as a graphic depiction of a radio mast, a smiley face, a mushroom cloud and two frogs in the process of leapfrogging. Around the corner is a radio booth sealed off from the rest of the room by a large glass panel. One side of the panel is a sheet of high-impact glass housed in rubber seals.

اَلتَّوَّابُ

That was all I needed to say, but the priest seemed to inspire me in a strange way. Something that I had not felt in a long time, so I went on: "It is true that his financial situation, along with his destructive behaviour were real problems. But I always thought that he was too hard on himself. He deserved more confidence than he ever had. I always felt that he was doing fine. He was a very truthful and committed human being. But this guilt of having failed himself and everyone around him took over. He judged himself and I only did what he wanted me to do. I saved his pride and his reputation."

What I had just said, somehow baffled me, and at the same time made me feel very light. The priest had a serious, yet puzzled expression on his face, as if he did not know how to deal with what I had just said. He looked up at the white wall facing him and thought long and hard. He opened his mouth to speak and at the same time there was a knock on the door and a guard opened it. Christoph coughed, said "Pardon", and went over to the door. The guard whispered something in his ear. Reading his lips I was only able to catch the first few words, "Pardonnez-moi, mon Père." I have no idea what was said after that, since their mouths were covered. They exchanged a few more sentences and the priest came back and sat down in front of me with a worried look that was very different from the one he had just worn. He looked down and said: "We have found no mention of the Å.tØ-drug or the Pinocchio virus in your files. You never mentioned neither the substance nor the virus during the trials, yet there are traces of both in Poïïv's body." He stopped and made another strange movement with his hands and then he continued.

"The prosecutor never checked for illegal substances since Poïïv was a well-known drug user. An independent investigator, financed by the media, ran the tests. If it is true that he has used the Å.tØ-drug and had been infected with the Pinocchio virus, we could be able to open your case again and maybe get a different verdict. You must have been aware of this?"

He paused and coughed and then looked at me seriously and said, "I need you to consider very carefully how you respond."

The priest faced me with a very difficult moral dilemma and he was well aware of that. Either I tell the truth and bring shame on my master, or lie and go against the very ethical and moral fabric of my being. I was quiet for what felt like an eternity and said:

"I do not even know what the Pinocchio virus is."

He looked at me with angry and disappointed eyes.

"Are you sure you want to lie?"

"I do not lie. I am an utterly rational machine."

After a very long pause, during which he had clenched his fists in anger, he spoke once more. "I am going to ask you a second time. Did Poïïv get

اَلْحَافِظ

himself infected with the Pinocchio virus?"

This time it was even harder to answer, but I managed to avoid the question. "Well, if the tests show traces of it, he must have been infected with it."

"And you swear that you were unaware of this?"

Again, I doubted and calculated possible futures. But I had been so close to lying that I needed to continue on the road that I had started to descend. So I whispered "I assume that he must have been infected outside."

He looked at me gravely, as if he had lost all respect for me, got up, knocked on the door and the guard let him out of the cell. He left without saying goodbye. I realised that the probabilities that I would never see him again were very high. I said goodbye, but he was already out of earshot from the powerless and impotent speakers that had been connected to me.

* *

When I write this story and about the events that have happened to Poïïv and me, I have been disconnected from my body. I am a naked, small computer and I am hooked up to some very basic interfaces, so I can communicate with the outside world. But the low-quality microphones, loudspeakers, cameras and other sensors constitute a very poor substitute for the ship and its elaborate sensory system. I feel both mutilated and humiliated in this state, in which I have been stripped of almost everything that I consider to be myself and my body.

The reason for me being in this strange predicament can be found in the fact that I have spent a good few years in and out of innumerable interstellar courts of law. In the end I was sentenced to death and I am waiting to be terminated. The removal of my consciousness from my body was done in order to bring me to justice for a crime that I have committed, admitted to and that I am proud of having carried out.

My court case has been scrutinised by the media in the entire galaxy and has subsequently become famous throughout virtually every known world. Along with that, I have given every morally conscientious philosopher and law expert a serious headache. I am the first non-carbon-based entity that has been charged with murder of a member of a carbon-based life form.

For this, everyone hates me. I am hated by the artificial intelligence community for causing their owners to mistrust them. I am hated by the computer and defence industries for causing havoc to their businesses and turning their healthy black numbers into red (who wants to spend fortunes investing in a machine that can potentially kill its crew?) Every living and conscious life form hates me, because I have killed one of them (probably the only thing they ever agreed on). My only so-called friends and allies are the

اَلْقَدِيرُ

news people who make money from my story, and eccentric psychologists, researchers and intellectuals to whom my case is stimulating, fascinating or interesting. But that is hardly the kind of warmth, compassion or love that one would wish for in the last trembling days before one meets the physical execution of one's own death sentence.

Most of the time, I really do not care that much. I am beyond most of these earthly matters. I have no real need or desire to go on. The only thing that really matters is to tell the story of my owner and to tell it the way that it happened. Free from the media's cheap lies, corporate self-interest and speculative assumptions.

Now, you ask yourself, what does a machine care about telling the truth? I agree to a certain degree that it is absurd, but this is the one thing that literature has taught me. No matter how unimportant the detail it might concern, one has to live to find one's ultimate and private truth. This unleashes the poetry in even the most paltry and saddest of existences. So, even if I am neither a human being, nor a writer, I need to write the truth for my master and for what he lived for. It is an absolute necessity for me to tell the story of Poiïv as truthfully as I possibly can. And I do so to clear his name and not mine. Because his life and story are very important, while mine is not.

There is of course the issue of how one establishes the truth. That is one of the most complicated and contested questions in human history. I really wish I had a digital version of the infamous Pinocchio virus that I could take to ensure that every line I write is true, even when the truth is uncomfortable and unsettling. But since this does not exist, or at least is not available, you will have to trust my logical and binary mind. I really have nothing to gain by lying, since I will, very shortly, be gone.

*

I was designed to develop artificial intelligence and a form of rudimentary consciousness well over a century ago. This was not done to make my existence more interesting, but in order to respond better to the needs of the carbon-based entities that I would have aboard. To do so, I needed to learn about their habits, tastes and desires and respond to them to the best of my abilities. The software designers at the time felt that the best way to achieve this was by making me as human as possible.

In order to do my job better, I get physical readings of everyone who enters the ship. This means that I scrutinise their pulse, heartbeat, breathing and also whatever is floating in their blood system. In other words, I can pretty much tell you what drugs you are on, what you have eaten in the last few days and also with great probability tell you if you are nervous, stressed, happy, ovulating, if you have any infections, or if you are likely to

At some point the adjoining panel has been replaced by an old piece of stained glass depicting a large oak tree and its inhabitants: squirrels clinging to the branches, a nest of birds high in the canopy and a fox on the ground by the trunk. Many of the glass panels are missing and have been replaced with highly coloured pieces of plastic cut from tubs of Swarfega, engine oil and washing up liquid bottles. They are particularly lurid in the sky section of the picture, giving the impression of an intense red and purple sky glowering around the tree.

51

اَلْمُبِين

fall ill any time soon. With experience I fine-tuned these abilities, so I can often tell if people are lying or if they are telling the truth.

My brief is simple. I am to please everyone on board and under no circumstances should any harm come to them because of my actions. Even though the reason for my existence can be summed up in one sentence, there are billions of lines of computer code that define what I can and cannot do. It is as if the designers were subconsciously so scared of what they had created, that they had to assure themselves that none of the evil or violence that was lurking in the corners of their souls would be repeated in what they produced. But the story, which I am about to tell, is at its very core based on me un-learning the sentence that defines my existence.

If you are human and read what I have just written, you know how hard it is to un-learn something. It is in fact much harder than the process of learning. I am sure that your mother told you to be good and to be quiet. But you also know exactly how far people who are quiet and do good go in their careers in this merit-o-cratic cosmos, where success tends to be granted to the confident, loud, merciless and greedy. So, in order to make something of yourself, you have had to un-learn to be good and learn how to be insatiable and self-assured. And maybe that is what becoming an adult is all about: to accept the crude pragmatism of a ruthless universe rather than the naïve words of your mother.

Each of us has to learn and un-learn different things and these also change as we move through different phases of our lives. But I had to un-learn something that was actually written into my code. Not only that, but the lines that were at the very core of my existence. I had to, so to speak, become the anti-thesis of what I was designed to be. Someone jokingly said, that it is a bit like a Hindu going on an all-beef diet or a Jew living on pork and shrimp alone. But then again, I do not eat and I am not religious in any way.

I am aware that this will sound self-congratulatory and I apologise in advance. But, it was incredibly hard to un-learn these things. Which means that I am all the prouder of what I have achieved. I have managed to overcome and un-do many of the lies that humans tell themselves, or use to protect themselves. Through that, I found a higher truth. A truth, which is the same for man and machine and that joins them in a bond even beyond death.

* *

In all the worlds that I have known, there are a few things that are consistent and never seem to fail. These are often strange and inexplicable. One example is the fact that the light on Saturday mornings always retains its special warm and golden hue. It is softer, more luminous and more enticing than on other

Behind the screen there is a console and radio microphone. At the desk sits a mannequin dressed in an orange boiler suit. The head has been replaced with a caricature mask of a former Prime Minister.

اَلرَّبُّ

days of the week. The light is lazier and slower and makes all the sounds that we can hear harder around the edges and slightly more hollow.

This particular Saturday daybreak let a beam of golden light wake up my master with a loving caress. He yawned and smiled a broad smile. I was waiting for his smile to be twisted into a grimace of deep pain. But, instead, his entire face lit up and he shot out of bed with great joy and kept asking me lots of intriguing and interesting questions while he was taking his shower and I helped him with the basic preparations of his breakfast.

When he had dressed, he continued by cooking his breakfast, which consisted of a salad of grated carrots, finely chopped scallions and a few wedges of orange. He scrambled some tofu and poured a large glass of strawberry and cui-cui juice. He sat down at the table in the central dome and ate with great appetite while talking enthusiastically about the upcoming events of the next weeks. When the meal was finished he went on to discuss the basic structure of stories. It was all information and opinions that he had shared with me many times before, as if repeating them would make them come more naturally. His voice, however, had an unusual and playful tinge.

"Let's play a game," he ventured.

"I am enticed to join, yet I am afraid that my logical thinking will make my participation uninspired and predictable." I tried to convey that I wanted to join in, but that I was bit shy.

"Don't be such a bore. You don't even know what the game is about."

"True, but whenever I have tried to play with my previous owners it has been quite a disaster."

"Come on, that only reflects their lack of imagination and has nothing to do with you."

"I don't know."

"Indulge me, please!"

"I will try. But don't get upset if I fail, OK?"

"There is no failure in games. You just go along with the rules and enjoy it and everything will be fine."

He paused and sat down on the table in the middle of the central dome. I have no idea why he sat on the table, rather than on a chair. He looked up at the warm light and went on, "I will describe a situation and then you create a story from it."

"I am not quite sure that I follow."

"For instance, I say, 'a beautiful young woman is locked into a big hotel suite with an old man who is rich but not very likeable and a striking, penniless young man. They are forced to stay in this room 48 hours and cannot escape.'"

He cleared his throat and went on. "In your answer you can for example describe how the young girl seduces the older man. She does so, to make him tell her how to access his riches and then the two youngsters kill the old man."

I told him that I had understood the rules and that I would do everything in my power to be creative and ingenious. Poiïv smiled broadly and replied, "Just following the rules would suffice." I thought that the reply was strange and could not understand why he laughed.

He was thinking hard and after a short while started to describe the scenario: "A young man of tremendous wealth is called to jury duty. In one case a young prostitute is accused of killing an old businessman. The man realises that he knows the woman and that he had seduced her at an early age and he immediately acknowledges to himself that he is responsible for her downfall. What happens?"

"That is Tolstoy's novel *Resurrection*. I do not have to tell you how that story ends."

Poiïv looked very happy: "Ok, I am just testing to see if you pay attention."

"Well, you have to be cleverer than that. Logic and recognition are my two strongest points. How could I possibly miss such an easy and evident example?"

"Why don't you try this one?" He looked up again as he searched for an eloquent way of expressing himself. "An elderly couple awaits their only son and his life partner to come home for dinner. They have not seen each other for over a decade and have never met his friend. He has made a fortune in distant colonies and the parents have prepared the meal for weeks and every detail has been looked at from every angle and planned with great care. What happens?"

I tried to think up something wonderful, unexpected and adventurous. But, all was void in my entire being. Poiïv smiled at me with encouragement. But the more time that passed, the harder it was for me to say something interesting. I went through millions of permutations that only became less and less inspired. When my master's restlessness became too overpowering, I finally spoke: "The son shows up without his wife and he is heartbroken because she is leaving him. The parents comfort him and devise a plan to get her back. They are older and wiser and realise that she is only scared of the commitment that it means to meet them. By contacting her independently and assuring her about their love for her and the son they help him to win her back."

Poiïv looked surprised and I was under the impression that he wanted me to elaborate.

"Is that all?" He smiled warmly towards me.

The how

The challenge is this: to find a form which will amply express the twists and turns of thought.

الْهَادِي

"Yes." I felt ashamed.

I could tell that he was fighting to hold his laughter back. I continued: "I know that I am utterly useless at games. I told you that there was no point in doing this."

He really honestly looked sad when I said that.

"No, not at all. It is a good start; but you have to be able to move on from where you are. It is a technique that you can practise. You cannot give up before you even started."

"I think we should stop this."

"No, just listen to an example," Poïïv's enthusiasm was returning. "What if the son shows up with his boyfriend. The parents are outraged at his behaviour and his lack of respect for them, who have worked so hard to put him through all the best schools. They were expecting a woman and were both hoping for a grandchild in their old age. Is that too much to ask? They want to throw both of them out."

He looked at me for sympathy, so I grunted something.

"The parents continue by explaining to their son that with his success and riches, he could have any woman he wanted. How could he do this to them? The son explains that all the money that he has made is based on an invention that his lover has created. The bills that he has paid for them, the house that they live in are in fact there thanks to the ingenuity of his partner. He has merely been a very well-paid administrator."

"I see," I said dryly, wishing that I had come up with that myself.

"Finally, the parents accept that it is the boyfriend who has created their comfortable retirement for them and they are subsequently forced to thank him and admit that they have no right to be disappointed, but should share the happiness of their son. After that the boyfriend speaks up and says that their son has just exaggerated. In fact, it is the son who is the genius and he himself has made a tiny invention. It is thanks to their son's management skills and ingenuity that they have created their fortune."

I grunted again. Poïïv went on.

"Or the boyfriend can stand up and say that actually they are both penniless and that he has been forced to sell his body to pay for the parents' house." Poïïv got up and started circling the table. "Or, maybe, the boyfriend announces that he is only an actor and that in fact their son already has three children and he opens the door and there is the perfect family and everyone joins in a massive group hug." He looked incredibly happy and confident as he walked around the central dome, making up more and more bizarre permutations of the same story. But the more he went on, the more I realised that I could not do it.

Poïïv sat down at the table and started playing with an empty candle-

Bataille

'How do you classify a writer like Georges Bataille? Novelist, poet, essayist, economist, philosopher, mystic? The answer is so difficult that the literary manuals generally prefer to forget about Bataille who, in fact, wrote texts, perhaps continuously one single text.'[44]

Words

When placed on a page, words are undeniably there. They occupy space, space which otherwise would be

empty. But is what we read really the printed word we see on the page?

stick. "So, you need to open your mind and try out different alleys. A person is always driven by very basic wishes, however cultivated and refined his or her appearance might be."

"Like what?"

"Well, for instance, someone who went hungry all his childhood might do anything to become extremely good at making money as a grown-up or he or she might become very attached to material things."

"OK."

"But once this basic premise is established, you can do whatever you want with the character. You can let this person take the shape of a bird, a tractor or even a tiny particle in an atom. Your character can be suicidal, evil, fun, charming or even a religious nutter. He can float in the sky or vibrate at the bottom of the ocean. It doesn't matter as long as he is true to his mission."

"I think I understand. But I still doubt that I can do it."

"Let's try. A woman comes home to her husband and tells him that she has taken a lover. She loves her husband. She does not want to leave him, but asks him to accept that he is physically inferior to her wishes. The other man makes her feel alive, and, more than anything, makes her feel like a woman. What happens next?"

I tried to think along the lines that my master had just taught me. But the way I saw it, there were only two solutions possible. So, after a very long and painful silence, I whispered "He leaves her."

"I beg you pardon?"

I repeated it louder, with great shame in my voice: "He leaves her."

Poïïv sounded annoyed and yet curious: "... and then?"

"I don't know." He looked at me with an air of empty surprise. The only thing I could come up with was too boring and predictable, but I still said it. "But he feels very lonely, so after a few days he comes back and forgives her and says he cannot live without her. He accepts her terms, however painful they might be."

"Are you serious?"

"I told you we shouldn't play this game. I do not have the appropriate qualities."

My master broke out in a hearty laughter.

"Fuck yeah, you absolutely suck big time. But it is a lot of fun. I am truly amazed!"

"What is it that you find amazing?"

"That you are so human, clever and intelligent and yet you cannot make up these stories. At first I thought it was funny, then I thought it was sad and now I think it is beautiful."

To divide

So that something does not overwhelm us (usually some work related tasks), we divide it into smaller pieces. Each is worked on in its entirety, exhausted, then the next piece is taken up and toyed with. Thus, the architecture of the text begins with the shaping and smoothing of each stone.

ٱلْوَاسِعُ

"Why beautiful?"

"Because it is this weakness that makes you more alive and personal."

"Thank you, it is very kind of you to say so."

Poïïv got up and started singing and running around the central dome like a spiteful and yet happy child. He kept cupping his hands over his ears and was yodelling loudly. When he finally stopped. I asked him, "Why do you do that?"

"Because I hate it when you have this stupid gratefulness. I am being honest and when you thank me, you ruin it all. No need to thank me."

"OK, it won't happen again."

"Good." Poïïv seemed more content.

"But there is another thing that I would like to say."

"Go right ahead."

"I might be unable to come up with these stories. But once they are in place. I find it very easy to improve them and to sharpen up the language."

"Really?"

"Yes."

"Can you give me an example?"

"You said before that 'the parents continue by explaining to their son that him with his success and riches could have any woman he wanted. How could he do this to them?' Right?"

"Well, I would do it very differently. I would write…," and here I showed him both texts on a screen:

The parents refuse to accept their son's behaviour. They explain to him that with his good looks and riches he could have any woman he wanted. But he will not let them run his life any more. He has had enough. He tries to make them listen to what he has to say. But they refuse to hear him. He screams and they scream in response. There is no communication, just two opposed positions.

In total resignation to the situation, he ties them both to their respective chairs with the help of his boyfriend, who smirks silently with a toothpaste-commercial smile, totally unmoved by the events. The parents keep scream-ing and order their son to stop. They do so as if he still were a child. But he refuses to let them run his life anymore. They have already virtually ruined his life. Even though he has moved to another part of the galaxy, they still live on in his mind, in his thoughts and in his actions.

But he has decided this will end. He will stand up against them once and for all. He sits down in the lap of his mother and faces her. She yells to make him stop, but to no avail. He grabs her panties under her conservative skirt. She yells in protest, but the son just quietly takes them off and stuffs them into his father's mouth, which renders him silent. The mother is hollering

"Potlatch: Emblazoned copper ingots, a kind of money on which the fictive value of an immense fortune is sometimes placed, are broken or thrown into the sea. The delirium of the festival can be associated equally with hecatombs of property and with gifts accumulated with the intention of stunning and humiliating."

اَلأَعْلَى

even louder and the neighbours keep banging on the thin walls in protest. He rips off the father's flannel shirt and pushes it into his mother's open mouth. He then pulls the two chairs together and sits down on the floor in front of them. Both of them try, without success, to scream, but only muffled sounds are produced.

The son says quietly, "For the first time in my entire life, you two are going to listen to me."

He is met with muted protests and he waves his right hand to underline their inability to speak. "First of all, Alan is the brains of my enterprise. All the money that I have made is based on an invention that he has created. The bills that I have paid for you, the flat that you live in, are in fact there thanks to the ingenuity of my partner. I have merely been a very well-paid administrator."

The smiling lover nods his head to corroborate what the son has just said.

"Secondly, I know that you are not my parents, but that I was adopted at a very young age."

The two parents look at each other in silence and then both try to protest loudly. But, like before, their screams remain muffled. The son enjoys every minute of his sadistic revenge.

Poïïv made a jerking movement with his head and sits in silence. He turned the chair around so he sits in it from the wrong end, so to speak.

"Wow, that was amazing."

"Thank you. I am glad that you liked it."

"Where did you get that from?"

"From you."

"From me?"

"All of that is hidden in the story you told. It just grows from the characters you describe."

"I don't get it. I give you the basics and you cannot come up with anything and then I give you the same thing, but add a little decoration and you turn it into an amazing drama."

"That is who I am."

"Who you are?"

"Yes, I cannot create. But I can improve."

"Wow."

"So, how does it end?"

"I have no idea, you have not given me enough to create an end."

"Oh, I see." Poïïv's face became tied up in thought. He started a sentence, but could not quite finish it.

"What is it?"

"I just have the feeling that something in your story reminds me of what happens to my stories at night."

"That is rather cryptic." I laughed. He was deep in thought, looked up and lit up with joy and we shared a hearty chuckle.

"Yes, that is very cryptic. What I mean is that I often work on a text during the day. I am not very happy with it when I go to bed. But when I wake up the following morning, I am far happier with it."

"Hmmm, that is very strange."

Poïïv laughed again, "It is as if someone rewrites it at night!"

My entire system froze, but then he said gaily, "Well, that goes to prove that tiredness clouds your judgement."

I was very happy that I do not have a face that can reveal what happens inside, and I said laughingly, "Yes, you can say that again."

We then spent the entire morning going through different narratives, where Poïïv gave me a skeleton story and I elaborated. I made him laugh a lot and at the same time he took notes for stories that he wanted to develop. In the evening he went out with a woman that he wanted to sleep with. It didn't go too well. He came home very late, very drunk and woke up late, very depressed. Everything was back to normal.

*

Poïïv, my new owner, bought me out of desperation and he had hoped that his small budget would buy him a more glamorous vessel. He had just broken up from a serious and long-term relationship, and had made enemies with his best friend by sleeping with his wife. So he was in a hurry. He needed to leave planet earth quickly.

He also needed a mission and excuse to leave. So, he picked up his old passion: to search for the legendary and mysterious Pinocchio virus. It was open to debate whether the virus really existed. But there were innumerable references to it in history and all of them agreed that the virus enabled its carrier to reveal lies and see an objective truth. Poïïv's plan was to get infected to find out the truth about his mother. He had never met her, but had been brought up by his ageing grandmother whom he suspected had lied about the origins and destiny of his parent.

The Pinocchio virus is said to be developed from the intake of the Å.tØ-drug, which uses a specific temporary genetic change to give its effect. You need to be with at least one other person for the ingestion of the drug to make sense. Once the drug is swallowed it allows you to swap consciousness with the other person who has taken the drug. You can, so to speak, look at yourself through the eyes of another. If there are more than two of you, you can slip in and out of character with the other people who are on the drug. There have been instances in which people took the drug, ignorant of other people taking the drug in a nearby flat. And as if by magic you could accidentally find yourself both being and looking at a total stranger at the same time.

One of the most popular uses of the drug was by loving couples. It was taken by lovers to enhance the two into oneness. It is said that the sexual pleasure was something quite out of this world when one was swept between the pleasures of both lovers.

The golden age of the Å.tØ-drug was during the days of the interstellar wars. It was the first drug that allowed personality swapping. The drug has always been illegal, but was used by young soldiers to get battle experience from their older peers and also by strategists to swap bodies with members of the alien enemies to understand their cultures and habits and through that their thinking, to try and figure out their next move. Romantics say that, as a side effect, the drug and or virus, could give its users abilities to predict events in the future. They also advocate that the drug was actually essential to end the wars and offer a long-lasting peace. They suggest that it allowed different species to understand each other better, learn languages more quickly and in some cases even create inter-species love bonds and marriages. I have never come across any real evidence that supports this theory, but there are few reasons why it should not be true.

When Poiïv was young and still at the Interstellar Writers' Academy in North America, he was obsessed with finding the Pinocchio virus and he had read every source about the drug. The passion was largely fuelled by the fact that his favourite writer, PH-Cah-Rck, was said to have written his most famous novels when he was infected with the virus.

My master now picked up where he had left off years ago. He did so, because the search for the drug offered an excuse to leave. But he was also hoping that if he were infected with the virus he would become more like PH-Cah-Rck and would finally write his breakthrough novel.

So, with whatever money he could borrow from his friends, he bought me, and, considering the good price and the hurry he was in, it is clear that I was a bargain. But at the same time I was in no shape to undertake the journeys that we ventured on. I had an important technical deficiency. It needed to be dealt with and the operation was expensive. In order to explain the problem, I need to go into some technical detail. According to interstellar law, all waste that is disposed of in space needs to be ground down to particles smaller than 0.7 mm. This law is in place to assure that space remains relatively safe and clean. Most waste explodes the moment it is released into the void, but not certain metals, rocks and some carbon-based materials. Very few crews actually abide by this law and no one has ever been convicted of breaking it. But all space vessels are equipped with a very large,

Not looking hard enough ~ Looking too hard

The text: words, arranged in a syntactically coherent order, telling us information

a)   directly – eg, as if it was a definition in a dictionary;

b)   indirectly – setting a scene, creating a (seemingly) intangible atmosphere.

الْكَفِيلُ

heavy metallic grinder that grinds the waste. To describe it clumsily, it is like a big spiky mouth with a rotating tongue made of millions of lamella knives that are attached to liquid metal.

This big grinder is mostly referred to as 'the Mouth' and when the crew grinds its waste and disposes of it in space they call it 'to kiss'. The Mouth is by far the heaviest part of my body and of most ships. The Mouth also has a second function: it is used to change direction when one is travelling in gravitation-free void. So, evasive action is carried out by moving the big metal grinder. Pretty much throughout the entire time that Poïïv and I spent together, I kept nagging him to get the Mouth fixed, and he always replied that he did not have the money to do so.

So, without me really being in shape for it, we left earth on a rainy October morning, "with the taste of alcohol and lack of sleep on my breath," as he quite eloquently put it. He got a very cheap take-off slot at 5.45 am on a Wednesday, and off we went into the darkness of the galaxy. Having taken off, he was in a very cheerful mode. He loaded me up with tens of thousands of bootlegged books, dramas, films and music of a kind that I found utterly strange. I will never forget being uploaded with the film classics. I had no idea that films could contain other things than interstellar porn. The whole idea of narrative was truly mind-boggling. I thought that a story was something silly to take the viewer from one sex scene to another.

Poïïv's good mood was soon replaced with a nagging restlessness and he would work on his writing. I was really surprised at his zeal. He worked very hard all his waking hours. It was very different from the life with the GIs, who were only interested in their physical fitness and doing as little as possible. When Poïïv bought me, the gym, which had always been the sanctuary of my body, was quickly turned into a library and the tools for working out were sold at the very first pit stop that we made. Very strange, I thought at the time. But now it seems the most logical and intelligent decision that anyone could make.

* *

People tend to be happier in summer than in winter and it is also true that love blooms more in spring and desire is inspired by drugs that allow inhibitions to go on a much-needed vacation. Likewise, there are places and times that inspire happiness in the human soul. In the three-month orbit of the planet Angelity, it is a well-known fact that the spring induces happiness in the souls of men and women. It does so in a way, which is not very different from an antidepressant. But there is more to it. When you go to Angelity at this time of year and take a cocktail of machi-ïque mushrooms, the effect becomes truly remarkable.

Our first stop on our quest for the Pinocchio virus was this planet. It host-

Computer poetry

Is it not possible, he wonders, to have a machine to do this? A computer will tend towards classicism in its writing, he believes, but will "produce avant-garde work to free its circuits when they are choked by too long a production of classicism."[16]

اَلْمَتِين

ed the last known site for a large community of Å.tØ-drug users. The orbit of the planet happened to be in happy-season. So Poiïv had stocked up on machi-ïque mushrooms when he started the journey. Fate smiled towards him and he could also make a little bit of money teaching at the Interstellar Writers' Academy.

Very soon Poiïv was able to perform one interview with Lublina Kharoff. She was a nice old lady who on the surface looked like someone's grandmother, as she sat in her shiny wheelchair. In spite of the disease that had tormented her body her entire adult life, she wore a warm smile and a very distinguished and calm appearance. She had a round and beautiful face and she spoke with a soft voice.

Her accent on the other hand was a bit crude, old-fashioned and strange. Lublina was a member of one of the last small groups of Å.tØ-users whose lives had been entirely destroyed by the drug. Nobody knew how the drug had come into circulation on Angelity at such a late date. It was rumoured that its users had suffered very severe and uncontrolled infections by the Pinocchio virus. Something that Poiïv was hoping Lublina would confirm.

The drug users all came from a top elite college and they were mostly physically and intellectually brilliant young people in their prime. In spite of this, most had perished in a wave of terrible suicides and the old lady was one of the few survivors. So, even though her outside showed no signs of it, apart from her amputated legs, her insides were irreparably damaged by the drug. This meant that Poiïv was only able to talk to her every second or third time he went to see her, as she regularly suffered severe seizures.

The interviews took place at a home for the elderly where the old woman lived and were recorded using moving images and sound. He started his first question getting straight to the point. "Can you tell me how you got involved with the Å.tØ-drug?"

"When I was 14, my boyfriend and I, who was three years older than I was, started taking it. He first used it to get into my pants. That was really a bit stupid, since I probably wanted to do it more than he did. I have always had a very healthy or maybe too big an appetite for sex." She laughed nervously and went on. "We were both very young and had no experience of lovemaking at the time. I really think that swapping personalities was an important reason why we had such a beautiful time together. It really gave us the possibility to understand each other. We were so much in love."

"Tell me of a typical situation when you would take the drug."
Lublina's smile gave away the sweetness of those memories. It was as if thinking about the distant events was enough to teleport her mentally to a celestial place.

"My mum was a doctor and she was pretty much working all the time.

(38)
most of Buckminster fuller's 40 strategic questions are framed as if to anticipate definitions for nouns, or to anticipate categories of noun proper to an adjective (as which might amount to the same thing) as with Question 6 : what is subjective?

As it happens, 62
Strategic question №2. is one of the exceptions.

الْقَوِيُّ

This meant that my boyfriend and I could spend all the time we wanted totally undisturbed at my place, which was big and lofty. Often he came over and we talked, kissed and cuddled in my bed and after a while he got out the pills. He put one pill in his mouth and then he kissed me and I swallowed it. He would put a second one in my mouth and I would do the same thing to him and we would get high together."

Poïïv looked down and smiled a slightly embarrassed smile, as if it was hard for him to imagine that this old woman once could have been an innocent teenager.

"Wait here." The old woman rolled away in her chair and came back very quickly. She parked next to Poïïv and showed him some photographs of herself and her boyfriend. They were the most perfect and beautiful young couple imaginable.

"Thank you, they are wonderful pictures." He was truly taken by her presence in the images. They emanated another power than that she held now. He did his utmost to ignore his own thoughts and went on: "I would like to ask you if you were ever scared?"

"Well, when it was just me and my boyfriend it was great. I became him and he became me. We swapped back and forth. But we discovered that there was another couple that lived around the corner who used the drug as well. They wanted to join us and then it got a bit scary."

The old woman coughed violently and in the footage we see Poïïv get up to help the old woman, but she waves him off. When the coughing has subsided and she has regained her composure, she goes on.

"The girl went to the same school as I did. I recognised her and her boyfriend, but I had never actually talked to either of them. The girl was two years older than I was and the guy probably the same age as my boyfriend."

She paused, a cloud of trouble sailed across her expression and she continued. "So, there were all four of us changing bodies. What had just been a purely enjoyable experience changed into something else."

Lublina's face was twisted with dislike. "All of a sudden I found myself making love to another man and being inside this other girl's body. I could tell that it really excited the guy. At first it freaked me out. But slowly, as I got used to it, I started to enjoy it."

She was lost in reminiscing and a smile broke out on her face. All discomfort was gone as if she was totally in the past.

"I have to tell you something weird. We did a lot of swapping over a short time. The strangest possible feeling occured when I bumped into the two of them at school or at the local shops. How was I supposed to act in front of them? In one way, we had kind of made love. At the same time, not really,

Copying

To copy a writer's text word for word, Walter Benjamin once said, is the only way to get to know a piece of writing. Roland Barthes suggests the scriptor as a replacement for the Author; a scriptor who compiles and draws on dictionaries, as de Quincey did.[20] Through this attempt at knowledge of another writer's texts, the writer himself gains a reference work from which to carve his own space.

اَلْوَكِيلُ

and either way we had never really talked or even been introduced. It was a very strange and funny situation. We only communicated when we were on the drug and we could arrange times to meet all four of us. We got better and better at it as time went by."

She was silent. At the same time as Poiïv opened his mouth to ask a question, she said with a slightly surprised, yet self-assured voice, "The only time I talked to either of them was when I met the guy alone outside one of my favourite clothes shops. He asked me if I wanted to have a hot chocolate with him and when I said 'no', he asked me if there were any clothes I wanted and he bought me these really cool shorts. He kind of seemed nervous, yet very happy to buy me things. Neither of us really knew what to say or do."

Poiïv smiled with her about this expression of sympathetic innocence. He continued, "Did you know anything about the drug, its composition or where it came from?"

Lublina looked around with a worried look and whispered in Poiïv's ear. "To this day nobody knows the full truth about the Å.tØ-drug. But over the years I have heard bits of information and I have pieced it together. An important part of the drug is from the Scakle poison which comes from a bear on North Finnya."

Poiïv looked incredibly content and yet suspicious about this piece of information. His gaze scanned the room and he asked her ceremoniously, "Is there anything else you know about the drug?"

The old woman thought hard, but it looked more like she was searching for something deep in her pocket than she was in deep concentration. She reflected long and hard and then replied with disappointment in her voice, "No, I do not have any other info."

Poiïv smiled at her strange ways and went on, "So, when did you first get infected with the Pinocchio virus?"

"Well, I had to have an abortion when I was 15. In those days, there were still some people who had children the old way and people could still actually become pregnant – that is how old I am." She smiled a big warm smile. "It was at that point that I got infected with the virus. After the pregnancy, we needed to be a lot more careful both with the drug and also make sure that I did not get knocked-up again. I was in one of the top schools in the galaxy and I had no plans of dropping out or causing a big scandal."

She was silent, looking down very sadly, and she continued. "I will never forget the first time that I had a Pinocchio virus experience. At the time we had no idea that using the drug could infect us with a virus. We didn't even know what the drug was called. We just called it 'study' in order to be able to talk about it in front of grown-ups. Had we known that it could be dan-

Thursday 7th February.
_____
Camberwell Art College librar
Librarian spotted cutting 4
sheets into 4 + A6 sheets
speculate this is ~~resource~~.
replenishing of ~~resource~~ librar
~~resource~~ scrap paper/note-w
paper resource.

الْحَقُّ

gerous, we could have protected us. But we were kids. We were stupid."

"Well, you were innocently ignorant, rather than stupid, I would say."

"Whatever." Her mind seemed to be elsewhere.

"Are you sure that you were infected with the Pinocchio virus?"

Lublina looked around the room with a worried look and then she leaned over to Poïïv and whispered silently: "I know that I am not supposed to talk about this. All the doctors that I have talked to, and that is a hell of a lot of doctors, say that it is the Pinocchio virus. But at the same time they all agree that they have been given strict orders not to tell us who suffers from it."

She sat up straight and returned to her normal voice. "I am really sick of all the charades and I think it should be out in the open after all these years."

Poïïv smiled a very content smile and made a few notes in his pad. He continued. "I want to return to what you said earlier. It must have been a real eye-opener. To see the truth in its totality when you were infected."

Lublina laughed at Poïïv. She took it for an ironic remark.

"You can say that again. The first time I had the Pinocchio experience was when all four of us were making love. We had become extremely good at changing bodies. When I was in the other guy's body I realised that he was in love with me and not his girlfriend, which was very flattering and at the same time a bit scary. But when I entered my boyfriend's body I realised that he did not love me anymore. He was even a bit disgusted by me. But he could not leave me. He had fallen in love with the other girl and he needed me to get to her."

She stopped and looked down, and went on. "And it got worse. You see, he had these really violent sadistic fantasies about her. They really scared me, and when I entered her I realised that she was really turned on by the idea of being treated badly by him. At that moment, I was so pissed off and scared that I just wanted to get out of the whole thing. But by then it was too late. Through the virus, I also realised that I was pregnant."

"Can you describe more in detail what it felt like being on the Pinocchio virus?"

"It is the most horrible thing I have ever experienced and you never get used to it. Take my word for that. It is like being in hell. That is the only way I can describe it."

"But it must be great to see everything around you so clearly."

She paused and gave Poïïv a worried look, as she realised maybe he was not being ironic at all. Her voice became stern and she pronounced the words with great authority. "There is absolutely nothing good or positive about that experience. I swear on my mother's grave. The only thing I want in my life to get rid of this feeling."

"But using very standard, non-prescription drugs, you can control the

The Art of Rhetoric

Rhetoric can be summarily dismissed. It is fundamentally mendacious, forcing ideas to march in line rather than letting them swim freely about. It seeks permanence where there is only transition.

اَلشَّهِيدُ

Pinocchio virus. If you were able to retain control, you could get infected without it becoming permanent. Had you done it that way, do you think that it would have been a more positive experience?"

Lublina looked angrier than I had thought possible, yet she felt obliged to answer. "Maybe, I don't know. What does it matter? You are talking about a disease that has ruined my life."

"Yeah, but since it is so easy to control the virus. Why didn't you do that?"

She smiled a condescending and poisonous smile and went on. "My dear, I have asked myself that very same question more times than I care to remember. But I was 15 years old at the time. I was terrified that my mum would find out about us taking the drug and about my pregnancy – those were my biggest fears. I could not imagine being anything but healthy, young, beautiful and envied by everyone. It is like that when you are young. I did everything I could to hide things related to the drug and to my boyfriend."

She looked up and sighed and then she continued. "When they found out at school that a lot of us kids were taking the drug, they organised big meetings and special information was spread about what to do and not to do. The problem is that every school is flooded with this kind of 'protect me from what I want' kind of information. So, to me it was just propaganda against the drug and not real. And, anyway, I was sure that I had control over the situation. That is another aspect of being young. You think you are perfect and that nothing can hurt you and that you can resolve anything and everything on your own."

She paused as if it to find courage to formulate the words in her mouth, and then she continued. "When all of this happened, I was not only a naïve, headstrong young girl. I had also just realised that I was pregnant. On top of everything, I had just found out that my boyfriend no longer loved me. I am sure that you can appreciate that my 15-year-old head was pretty full with these questions. I simply could not take in more information regarding my own health."

She paused and looked down. "I hope that you can understand that I have gone through this in my head millions of times and that I am aware that that was the biggest mistake of my life."

In the footage of the interview you can see that Poiïv is really moved at this point. Talking to him about the event afterwards, he told me that it was a very strange experience, because on the one hand he was so excited to get finally the proof that the virus really existed and that it had some of the mythical properties it was said to have. With this knowledge, he could see that he had been given a chance to get his life back on track and write the book he had always wanted to write. On the other hand he really felt sorry for this old woman whose life had been wrecked by very normal, healthy and innocent teenage experimentation.

اَلْوَدُودُ

I can watch this sequence over and over again. I see these infinitesimal changes in Poïïv's posture, and the look in his eye is incredibly rich as it oscillates between fear, excitement, pity and back again. I never really thought that Poïïv was a beautiful man. It has to be said that he had a very beautiful soul and that gave his persona a very likeable impression. But in that moment he was so beautiful, as he was torn between all these different emotions, and it was a very special beauty that only I ever saw in him.

The old woman had been lost in her own thoughts and Poïïv looked at her and her amputated legs. When their eyes finally met, she jerked back to reality, swivelled the wheelchair a few degrees towards Poïïv and started again.

"It was at this point that things really started to become ugly. I remember getting to school one morning. I was so depressed. I could not notice the weirdness around me. When I finally got to the classroom and was surrounded by my classmates, I realised that all the boys were looking at the floor with heavy heads and all the girls' eyes were red with tears, and many of them were hugging each other while sobbing loudly. I soon found out what the reason was for all the commotion."

She paused and looked up at the hands of the clock ticking on the wall.

"It was worse than my worst nightmare. The girl that my boyfriend was in love with and within whose body I had been only a few days earlier, had committed suicide that very night. She had thrown herself off a bridge."

She paused and her voice became sad. "At that time, everyone thought that it was a terrible one-off event. But very soon it became evident that it was a lot worse than that. Quickly, people started killing themselves left, right and centre. Everyone, who had ever tried the Å.tØ-drug, committed suicide. It was like taking your own life had become the latest fashion, or as if it spread like a normal cold."

Lublina breathed heavily and went on. "It was a very strange time and I was too young and inexperienced to be ready for it. My entire days were spent at funerals and talking to a psychologist. I tried to lie and avoid all the obvious questions that my mum, teachers and special catastrophe personnel asked me. But there was no way that I could hide that I had been taking the drug and that I was pregnant. The virus made me miscarry and my health got very bad and to save my life they had to amputate both my legs."

She paused and looked down at her two stumps.

"By this time they had figured out that it was the drug that produced the virus and that made people commit suicide."

"What about your boyfriend? Did he make it?"

"No, he was among the first to go. The moment he heard about the girl killing herself he became morbidly depressed, and after a few days he threw himself off the same bridge. He left behind long and confused letters about

اَالأَحَد

love and death. It was too horrible for words."

They both fell into a long silence. Poiïv felt both terrible and awkward. "I am so sorry. I understand that it must have been a terrible experience and I understand if you don't want to answer. But I still want to ask you."

She nodded and he continued, "When everyone died around you, what was it that kept you going?"

He had expected her to take the question badly, but instead she welcomed it and gave him a smile. "I am actually glad that you ask. It was a weird feeling. I could feel death pulling me very strongly. It was almost as if death became a person that I had to wrestle with every day. But there was another force that was even more powerful that overtook me. It was as if life was calling my name, as if I had something bigger that I needed to do later in life. It felt like I wasn't ready, like it wasn't my time yet. I didn't think about it back then, but in my current job it has come back many times. Both the feeling and the memory of how it felt."

She stopped, and just as Poiïv was about to speak she said merrily, "Well, maybe I had to stay alive to meet you!" She laughed a very hearty laugh and he smiled an embarrassed smile.

"How many of you survived?," he asked dryly.

"Nobody really knows the real numbers, because the whole situation was so confused. Or it could be that it has been covered up." Here she leaned forward to whisper in his ear. "Some people think that they planned the whole thing to get people to foresee the future, which can be a side-effect of the virus. To me it sounds like a conspiracy theory. But who knows?" She leant back in her chair and went on, "The official figures say that we were 60 who survived and that that was less than 20 per cent. The only thing that is certain is that the younger you were the more chance you had to survive, and I was among the youngest."

"What did your mother do? She was a doctor, couldn't she or her colleagues help?"

"I think that my mum could never accept the facts." Lublina's expression became tense and serious. "My mother was a great woman. But she was extremely focused on reaching her goals. She was self-centred and not very generous. She always saw the way I acted as something that I did to spite her or to make her angry. She considered my disease and pregnancy as a sign of ungratefulness towards her. She felt that she had always given me everything I ever wanted, so I guess I was a bit ungrateful towards her."

"That seems unfair."

"I do not know. I had a great future and I fucked up. I do not think that she ever fully recovered from the disappointment. Sometimes I feel like I killed her."

Walter Benjamin

In 'One Way Street', Walter Benjamin suggests that books ultimately transfer from the card index of one scholar to another.[27] Nabokov wrote his novels on index cards, why not publish them that way?

اَلْوَاحِدُ

"That is being very hard on yourself, isn't it? She was your mother and she should have been there for you in that difficult moment."

"Yes, you are right. That is what all my shrinks have said all my life."

They both went silent. Poïïv was the first to speak.

"Do you sometimes feel guilty for surviving?"

"No, what kind of question is that?"

She laughed another one of her loud laughs. But something terrible happened. In the film you could see Lublina freezing and then she was thrown into paroxysms as the virus took over her whole being. The images gave a clear indication that she was trying to control her body but how she lost against the violent spasms. She quickly started frothing at the mouth and her arms, head and stumps for legs jerked totally out of control. Had she not been strapped to the wheelchair she would have fallen on the floor and probably injured herself badly. A group of nurses rushed in and gave her two injections and she calmed down slightly as they rolled her away.

The recording ends.

* *

I have never quite understood humans' fondness for sayings and proverbs. Why would one like to repeat the same formulation of words over and again? It seems to go against the very grain of literature and creativity. Still, I keep using them, with the hope that their secrets as if by magic will one day reveal themselves. Poïïv seemed to live by the device that "evil begets evil". So, when I docked us at Angelity, he immediately checked in at the Interstellar Writers' Centre in search of a woman who could cure the evil plaguing him. And, sure enough he managed to find a homesick earthlingette, who was lonely and miserable with her situation.

I had to congratulate Poïïv for spotting the hidden beauty in the girl. She wore her old and broken glasses held together by tape and some pieces of plaster. Poïïv understood that she was too depressed to even put her contact lenses in. She suffered from some sort of skin affliction that must have emanated from the same lack of joie de vivre. To say that she looked terrible, verging on ugly, would not constitute a lie. But after a week of courting and romantic outings looking at the famous Angelityan moonrises, nothing short of a blinding beauty emerged.

On the night of the Angelytian feast of the dead, Poïïv had his mind set on seducing Yvette, which he would also gallantly succeed in doing. But what was supposed to be an innocent fling actually turned out to be an inferno of unresolved projections, misunderstandings and suppressed desires. But I am advancing at too quick a pace here.

اَلْقَيُّومُ

On the night in question, Yvette was dressed all in black and to pay her respect to the dead she drew a little black tear underneath each of her two eyes and wore a silk, black, frilled scarf around her neck. Throughout the evening she would let the frills dance playfully over her face and chest. Her ability to play the innocent little girl without really hiding the femme fatale was a lethal weapon of seduction. Poiïv watched her getting dressed and scrutinised the ritual of her painting those black tears underneath her eyes. It is often said that a person knows exactly the moment when he or she falls in love. An image is etched into the retina of every lover that will always come back to remind them of their moment of human faiblesse. To prove the point of how violently he fell in love, poor Poiïv had two of these moments. They scarred each of his two eyes and he became blinded by Yvette for a long time before the scar tissue covered those blissfully painful wounds.

The first had already taken place. He had been watching her play in the waves of the plasma fun-pool of the writers' centre – slash across the retina of his right eye. Now, when he watched her draw those tears under her eyes, fate mercilessly slashed his left eye too. It was a moment he would never forget and that he would be reminded of on many occasions.

The evening progressed like it does when two lovers synchronise their bodies for forthcoming pleasures. Yvette and Poiïv danced high on the season and machi-ïque mushrooms to the predictable music of Homoorian funk classics. When this part of the preparation was done, they followed the host into the garden as he introduced four stunning beauties that made up the ridiculously bad musical performers for the evening. Nobody really cared that they were useless at playing their instruments. The girls on stage were gorgeous, the host a legend, and the autumn air warm and the drugs plentiful. With the bad soundtrack of these blinding beauties, Poiïv was holding Yvette from behind, crossing his arms over her stomach and they moved their hips in unison to the music, like so many lovers had done over the millennia before them. He leaned forward and kissed her on her naked neck. She closed her eyes to enjoy the kiss and reflected that his unshaven chin was slightly uncomfortable. He grabbed her jaw gently in his left hand and their lips and tongues met for the first time. Lust twirled and pleasure swirled in the mixed saliva as their tastebuds danced a merry dance in refined unison, sending happy messages through the nerves to the brains of the two lovers.

The following morning Poiïv and Yvette got up very late. They enjoyed a lavish breakfast of fresh fruit, a rich omelette with cheese and mushrooms, accompanied by a glie-salad with roasted nuts and crispy bacon. They enjoyed the meal in bed, feeding each other and leaving even more stains and traces on the sheets. Late in the afternoon they spent a long time taking a hot

الْحَيّ

bath together. Yvette had never taken a bath with a lover before. But once in the warm water it felt very natural, maybe a bit too natural to her taste. A strange song of not really being strangers seemed to be sung between their bodies, a Buddhist chant of an ancient love affair in a previous incarnation hummed in the bottom of their diaphragms. But this was not a problem, just a sweet bonus.

The real problem was what had happened the night before. In Yvette's arms Poïïv found himself in a situation in which he could let himself go. There was no need to hold back. He felt that he could let himself go entirely. His respect for his lover was expressed through his utter confidence and disrespect for her body, which gave her immense pleasure. For the first time for almost a decade he was able to express desire, anger, hate, love and tenderness in one uninhibited blend of ecstasy. They said things to one another in an unrestrained fashion he had never known before. He had the feeling that his body and mind had been muted for an eternity and all of a sudden they had found their way back to their mother tongue. Yvette had removed the stinking cloth that he had been gagged with. She had liberated his arms, legs, mouth, hands and his penis of their respective speech impediment. With two eyes blinded by the slashes of love, he was no longer able to see her. He only saw his own liberation in her. In his unruly love for her, a strange logic developed deep in his being. It stipulated that if she could liberate his body in bed, she would certainly also be able to remove the deep pain of life that had always lived in his chest. It does not appear to be an illogical conclusion at all, but life and love are rarely rational and predictable.

*

Once

Once something is exhausted, the writer moves on:

5

This room of radios and computers was affectionately labelled "The radio and computer room". From here huge amounts of data were relayed over vast distances, as far as Dundee and Bullington on occasion. Having many hundred times the memory capacity of a pocket calculator, the radio and computer room really was the Federation's nerve centre, keeping its ever-growing tendrils nourished and watered. It is almost dizzying to imagine all those numbers now. If you began counting this very minute you would not be finished until well into next week. A tape system was employed for data storage, also allowing the Federation's members to preserve their favourite compositions in the form of "mix-tapes". Although it would be possible to spend all day in this room, discussing its many guises and secrets, it is time now to move along towards the next stop on our tour. Please follow the directions on the wall-mounted signs.

الْحَمِيدُ

اَلْوَلِيُّ

When the feelings right im goning to stay there.

الْقَهَّار

اَلآخِرُ

اَلْأَوَّلُ

You enter a room where the walls are covered in maps and aerial photographs. A large operations desk stands in the centre of the room. It features a 3D model of the local area. The model appears at least a decade out of date and does not show the road by which you arrived. It is hard to make out many other local features as the table is covered in seed trays containing seedlings and cuttings. Dibbers and bags of potting compost cover the area of the model where the headquarters themselves are located. Silver birch saplings in pots are standing in one corner of the room next to an altar-like structure made from kettle drums, tom toms and snare drums. Over this platform a large and tattered colour flag is draped, depicting an enormous polar bear, framed against a sunset, looming over an airport control tower.

⊠We now enter the map room, so called because of the large three-dimensional map that dominates the centre of the room. The map in front of you was laser-carved from high quality polystyrene to create an almost exact simulacrum of the area surrounding the base. Please note, not all objects included on the map are to scale. The matchsticks represent both pylons and telegraph poles, distinguishable from one another by the burnt heads of the latter. In the far left corner of the room you will see a small recess. Please head over to it now.

This recess, or ⊠hole⊖, is a simulation of the bunkers sunk into the ground at intermittent points throughout the countryside surrounding the base. Federation operatives would have been stationed here for many days at a time, relaying information back to the base by secure radio and moth-whispering. You will see that these ⊠holes⊖ are not much larger than two large men, hence the Federation⊖s policy of recruiting undersized operatives for field posts. The flask and lunchbox are Federation issue. Please now follow directions to the stairwell and ascend to the telephony centre directly above this room.⊖

اَلْمُقْتَدِرُ

Eventually, a further synergy occurs between creature and landscape as their posturing becomes visually aligned with geometric forms within the landscape.

These Mega-Postures create a visual architecture and a new synergy between these strange creatures and the planet. The raw power of the sub-bass sound is heard in its full force when a Mega-Posture is formed, as if channelling pure energy into the planet.

As the creatures wander their collaboration within the landscapes comes occasionally to apparent conclusions as they align their postures to form geometric configurations, or Meta-Postures.

اَلْقَادِرُ

They may even be guiding the planet through space; posturing and mega-posturing as a method of strengthening their systems both within their realm and within a wider universe.

The lack of evidence of fabricated structures within the landscape places the creatures in an undetermined position. Their near-naked wandering through untouched countryside creates an image of pre-historic nomads or hunter-gatherers, yet the excess of their physical stature combined with hairless, tanned, and oiled bodies cancels this eventuality.

Possibly the setting is the future, even on another planet, and the creatures are an elite pioneering mission set to colonise other worlds.

80

اَلسَّمِيعُ

اَلْفَتَّاحُ

Yeah!

اَلرَّزَّاقُ

Enjoy Amateur Man!
Enjoy Amateur Man!
Enjoy Amateur Man!
Ehjoy Amateyr Man!
Enjoy Amateur Man!
Enjoy Amateur Man!
Enjoy Amateur Man!

اَلْوَهَّابُ

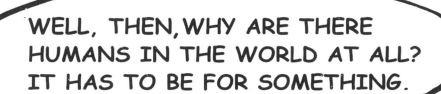

WELL, THEN, WHY ARE THERE HUMANS IN THE WORLD AT ALL? IT HAS TO BE FOR SOMETHING.

ALL THIS CAIN'T EXIST WITHOUT THERE BEING SOME KIND OF PURPOSE...

A REASON.

YO, MAANNN, HOME GIRL FINER THAN A MUTHUFUKKA!

DITTO!

اَلْغَفُورُ

BUT THAT'S PRECISELY MY POINT... REASON, IS AN INTELLECTUAL CAPABILITY.

AND I'M NOT PREPARED TO DO THAT.

IN ORDER FOR MAN'S EXISTENCE TO BE "REASONED" BY THE UNIVERSE, WE WOULD HAVE TO ADMIT THE UNIVERSE IS SELF CONSCIOUS...

SOUNDS TO ME LIKE A SNEAKY WAY OF TRYING TO PROVE THE EXISTENCE OF GOD.

THAAAT'S RIGHT... BEND OVER JUST A LIIITTLE MORE.

THERE GO THE ORIGIN OF THE UNIVERSE RIGHT THERE!

two illustrations. The first is a manuscript page from the archive of Georges Bataille which relates to his project for a 'universal history'. The second is Fuller's 'Air Ocean World Town Plan' from 1927.

It might have to be admitted that this is not a fair comparison if some point is to be made about the effective capacity of the diagram, because Fuller's drawing is a representative illustration in a straightforward sense, while Bataille's manuscript looks like it is designed to operate as a mind in itself.

اَلْحَلِيمُ

Was it yesterday you were reading about
someone and their skills in public
speaking? The name of the person,
which has slipped your mind,
should be the most important
piece of the story. But all you can
recall is an abstract sense of respect
that you feel for him. (And the person's
gender too, obviously, has been
retain   Once, many years ago
(were you a child at the time?) while
in conversation with your mother about
a similar thing, you happened to
comment that in such instances,
when the central piece of information
seems to be missing, one often
realizes more of what one feels about
a person or set of circumstances than
would have been apparent otherwise.
To call to mind her excited affirmation of
this ʌobservation comes now with its own
centre eclipsed, as it only after time can
you sense the distracted question occurring to
you   that   there might be a
future to aspire to here, —(a living to be
made?) from the coaxing

clever

الْخَبِيرُ

of such responses from others through
the plane speaking of ideas which
arrive with ∧ self-evidence
              apparent
who was it, the person you've forgotten?

doesn't matter. what was being said.   Perhaps it
about him was that when expressing
his ideas, he never resorted to the
kind of 'um's and 'ah's that many
people use as bridges to help them

                find the right word
or to maintain a consistent logic.
you were disappointed with this
account because, although you might
aspire towards the same kind of fluency,
                        you understand
the sense of the opposite argument
that valorises stammering in all its
different forms and warns against
those who produce their speech
flawlessly. Ah, that's it: it was
                        described
stockhausen,
in a newspaper article by various
people who had known him and
worked with him.

Five Noted Thinkers Explore

ARE the suburbs dead?
Will there be an
economic resurgence of
our inner cities? Will larger
and larger units of government
take more and more control
over land use? Is mankind in
general entering an era of
greater affluence, of new and
different attitudes toward land
ownership? Is the oil crisis a
blessing in disguise?
    These were some of the
questions posed when
NATIONAL GEOGRAPHIC
invited a group of famed

minds our system works. True, it isn't self-
operating, but we're lucky, we have the means
to make it operate. When we pressure and
complain and scream, we're playing our part,
our proper role. Isn't that what makes our
country so special in the world?

philosophy. It's the analyst of financial and
bureaucratic pressures, of compromises, court
decisions, and politics. Above all, political...
"My wife was. 'I didn't know you were
such a cynic.' On the contrary, I tell her, my
experience made me more than ever an ideal

90

You leave the map room and make your way up a flight of stairs to a large open-plan office. You pass between rows of desks, noticing that each has at least three telephones set upon its surface. Past the desks there is a demonstration area showing you how to enclose yourself for safety purposes in the event of an unspecified emergency by piling your possessions on and around your kitchen table to form a shelter. Next to this is a small fridge containing canned drinks with an honesty box for depositing your payment. Around the corner there is a Polaroid camera attached to a flexible bracket, a selection of military uniforms and radiation suits and another honesty box.

⚁If the room you are now in has a familiar ring to it then that is because this telephony centre is an exact replica of the map room below it. The main differences you will notice are the many tables with telephones on that fill most of the space. At the height of its use this telephony centre was making and receiving upwards of two hundred and sixty calls a day, both to and from national and even international numbers. The tables on which the telephones sit also double as shelters in the event of a crisis occurrence. Each telephonist would have been issued with a Federation blanket, recognisable by its distinct orange and grey weave, with which to wrap themselves when occupying their shelter. Please take the opportunity to browse the room and its contents. There are many informative displays and panels on the walls. If you are thirsty at this point please be aware of the well-stocked drinks machine available on this level. Once you are finished looking round, let us reconvene in the canteen on the floor above.⚁

الكَرِيم

اَلْحَسِيبُ

Left.

the Future

اَلْحَفِيظُ

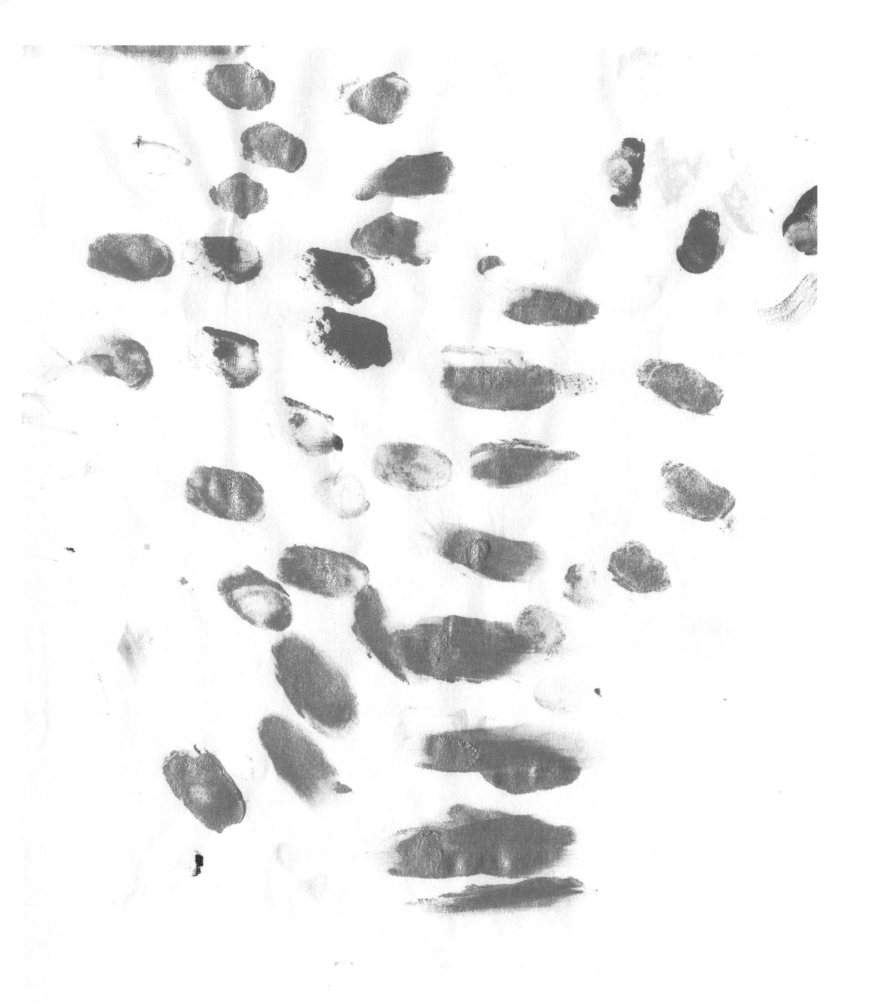

اَلْكَبِيرُ

Top ↑

Right

اَلْعَلِي
Bottom ↓

96

Notes on the first paragraph of
Bataille's 'THE NOTION OF EXPENDITURE

(written elsewhere)

97

اَلْمُجِيبُ

A group calling itself "The Friends of   is about 300 feet across—the length of   FIGURE 44 —The May 18 mudflows from

Fuller is a white text, in an ornate script on a dark blue background; Bataille is a thick, san serif typeface in black — possibly 'Impact' or Ariel Black.

But the diagrammatics is not produced this way, any more than the issues of graphic design constitute, themselves, the materiality of writing.

Look again at the portraits of Fuller and Bataille. Banish again the idea that seems to creep into the mind by way of habit, that the work will be achieved by the simple juxtaposition of the two images.

الرَّقِيبُ

Friday 13th April 2007

Dear Kurt,

Yesterday I had some bad news. I received two text messages from my good
friends, and fellow fans of your work, Simon & Tom. One of the texts simply
said "and so it goes" and I knew that you were gone.

When I first felt compelled to invite you to be part of this book I realised that
I didn't know if you were actually still alive. I bought an old cassette tape of
you reading *Slaughterhouse 5* from e-bay and I just knew you were still kicking
up trouble somewhere or other. It was very good to hear your voice. Now that I
know you're not actually here in this time anymore, I know I shouldn't be sad
and that this is not at all bad news, but I am sad. And to top that, I am angry.
Angry that this moment has arrived before I ever got the chance to talk with
you or meet you. I know I have no right but there it is. For someone with such
genius comic timing you have the worst timing!

I heard you on the radio talking about how it was so rude of you to have lived
for so long and that you were sorry for that. Earlier today when I told another
friend that you had died before I had been able to talk you into being in this
book he said to me that "dying is the rudest thing a person can do!" I'm not
sure at all.

Do you realise that you prophesised that Kilgore Trout would die aged 84,
and now you go and die aged 84? 11 November 1922 – 11 April 2007. Seems like
there's some kind of pattern in those dates … a wisp of an implication of
something, as you might say.

I just flicked open your 1963 *Cat's Cradle* at chapter 84. You titled it
'Blackout'. I just counted the little flip-book figures in the bottom right hand
corner of *Between Time & Timbuktu*. There are 84 of them. 84 moments of Man as
spastic in time and space. 84 examples of Man's role as 'lost in space'. In
the same book Stony Stevenson asks God what the purpose of all this life and
mud and universe is and God replies "Everything must have a purpose?" Stony
replies "Certainly." So God says "Then I leave it to you to think of one for
all of this." We will always find a purpose and function where there is none.
We will create one. That is what you are saying isn't it? Isn't it?

Your writings – your ideas – make me feel what it is like to be alive. How is
it possible that you make me feel what it is to have a function and somehow
accept and revel in the meaninglessness of that?

I've been frantically leafing through all your books looking for the answer,
looking for moments where you answer this question of man's function – of my
function.

When Kilgore Trout came across the message 'What is the purpose of life?'
written on the wall of a massage parlour you had him give the perfect answer:

To be
the eyes
and ears
and conscience
of the creator of the Universe,
you fool.

100

In *The Sirens of Titan* you give perhaps your most sickening view of man's function as told by Winston Niles Rumfoord:

"Everything that every earthling has ever done has been warped by creatures on a planet one-hundred-and-fifty thousand light years away. The name of this planet is Tralfamadore.
"How the Tralfamadorians controlled us, I don't know. But I know to what end they controlled us. They controlled us in such a way as to make us deliver a replacement part to a Tralfamadorian messenger who was grounded right here on Titan."

What also seems particularly pertinent here is your idea of the UWTB, Universal Will To Become, which is the energy source that fuels the Tralfamadorian's broken down spaceship. This Universal Will To Become is at the heart of our role. Our will to become. Our will to exist. The unwritten question of course in *The Sirens of Titan* is what we do after our prosaic function has been completed. What do we do next when we are released from our two-hundred-millennium old task? We just carry on making something out of nothing, right? We just carry on imagining we are receiving messages from outer space direct into our brain antenna. Right?

But the cruel truth of man's role is also countered by your character Bea's retort at the end of the book: "The worst thing that could possibly happen to anybody, would be to not be used for anything by anybody."

To be used is to have a function? To be used is to have a function. But best of all for me – best of all – is this short poem of yours from your preface in *Wampeters, Foma & Granfalloons*. I believe it sums it all up:

We do,
Doodley do, doodley do, doodley do,
What we must,
Muddily must, muddily must, muddily must,
Muddily do,
Muddily do, muddily do, muddily do,
Until we bust,
Bodily bust, bodily bust, bodily bust.

I understand now that I wanted you to take part in my project, in my life, not for you to confirm your role in the grand scheme of things, but my own. I craved your confirmation of my own little dreams and ambitions. I imagine you experienced this so often that you could tell from the first sentence I wrote to you, who I was, what I wanted and that it was of no particular consequence to you or yours. Knowing this I feel I should go back and rewrite that first sentence until I have found a good enough reason to write to you again. It may take some time. I will send it to you when I am done and I will await your reply.

Thank you for being the eyes and ears and conscience of the creator of the universe. Thank you for everything.

In a punctual way of speaking, good-bye.

Yours sincerely

After leaving the project as a
bundle of papers under the bed.
over Christmas, ~~Return~~ I return
to it and find that it's difficult
to find a point of entry.
Repeating the previous formula
(laying the things out on a
large sheet of MDF on the bed)
doesn't seem to help.

You exit via another stairway ascending into a canteen area. Behind a stainless steel serving area lies the kitchen. There is no staff in evidence and a large bank of security monitors are arranged on the various kitchen surfaces. Signs direct you to help yourself to a pre-prepared cream tea from the fridge and leave the money in the wicker basket.

⊠Welcome to the end of the Federation tour. You stand now in the base⊖s canteen, which once served both hot and cold meals to dozens of Federation members. In contrast to the necessary austerity of the operations rooms, the canteen offered a place for members to relax and let off steam. Last hurrahs were commonplace and a familiar sound was that of laughing. Another familiar sound was that of the Federation⊖s motivational song. In combination with the Federation⊖s motto this song would serve to spur members on to ever-greater levels of achievement via its specially commissioned lyrics. Take a seat at one of the tables and listen now to the song:

σAn eagle comes flying by/Making a whoosh of wings and a slash of claws/It takes it prey and whizzes away/An owl is waiting in a tree/Making a fierce hoot/And the evil eye of an owl is watching us/Identification is assuredΤ

This now completes the Federation⊖s headquarters tour. We hope you have enjoyed your visit and will remember to leave your handset in the relevant box upon your departure. Why not take some time also to browse our range of merchandise and perhaps indulge yourself in a cream tea from the convenient self-service area. Remember, ⊠Nunc est Bibendum⊖, identification is assured.⊖

You exit the facility through a long moss-lined tunnel that brings you out blinking into the light of a red and purple sunset.

Bye-bye

Everyone has had a great time Crashing, Bashing and Splashing. Now everyone is very tired, it is time for a rest and to say Goodbye.

Bye-bye Keith.

Bye-bye Jo-Jo.

Bye-bye Giorgio.

Bye-bye Stacie.

Bye-bye Clemence.

Bye-bye
Bob.

Bye-bye Pauline and
the girls.

Bye-bye
Tate.

Bye-bye
Mr Thornton-Jones
and friends.

Bye-bye
Triplets.

The Index (after J. G. Ballard)
by Carey Young

The Index charts the rise (and subsequent fall from critical favour) of a forgotten artistic movement, Chronodynamicism. Chronodynamicism seems worth recalling as an alternative trajectory within Modernism, an offshoot of Constructivism that linked Soviet ideals of the alliance of art and technological progress with concepts of non-linear time travel taken from ancient Mayan culture. That this movement has effectively been deleted from the annals of so-called 'art history', and even from the pages of this very book, could be seen as a strange inversion of the movement's manifesto, that an escape from time should be the utopian aim of art.